American Institute of Chemical Engineers

Catalysis and Reaction Engineering Division Conference

Presentations at the
2008 AIChE Spring National Meeting

April 6-10, 2008
New Orleans, Louisiana, USA

Printed from e-media with permission by:

Curran Associates, Inc.
57 Morehouse Lane
Red Hook, NY 12571
www.proceedings.com

ISBN: 978-1-60560-223-3

Some format issues inherent in the e-media version may also appear in this print version.

ISBN: 978-1-60560-223-3

American Institute of Chemical Engineers
Proceedings
Three Park Avenue
New York, NY 10016-5991
Phone: 212-591-8100

www.aiche.org

American Institute of Chemical Engineers

Catalysis and Reaction Engineering Division Conference
2008

TABLE OF CONTENTS

Enzymatic Synthesis of Small Chiral Alcohols in One-Phase and Two-Phase Systems 1
Pia Müller, Lasse Greiner

Hydroformylation of Higher Olefins with Immobilized Catalysts 2
Timo J. Koch, Jürgen Klankermayer, Walter Leitner

"Pot-in-Pot" Reactions: Polymer Macro-Fabrication for Site-Isolation of Catalysts and Reagents for Cascade Reactions 3
Martin T. Mwangi, M. Brett Runge, Ned B. Bowden

Reaction Calorimetry as Means for Process Intensification of the Epoxidation of Methyl Oleate 4
Florian Pontzen, Daniela H. Mueller, Marcel A. Liauw, Lasse Greiner

Rapid Lauroylation of Wheat Straw Hemicellulosesunder Microwave Heating 12
Junli Ren, Feng Xu, Runcang Sun

Highly Stable Pt-BHA Nanocomposite Catalysts for Low-Temperature CO Oxidation 13
Yi Zhang, Sittichai Natesakawat, Tom Sanders, Christopher Matranga, Goetz Veser

Kinetic Modelling of *n*-Dodecane Dehydrogenation on Pt-Sn/Al_2O_3 Catalyst Based on LHHW Mechanism 14
Parisa Valipour, Mahdi Alavi Amlashi, Soosan Rowshanzamir

Aerobic Epoxidation of Alkenes with Manganese (III) Salen Complexes Immobilised Onto Aminopropyl-HMS 26
Jamil Khan, Peter Styring, Adam F. Lee

Utilisation of Heterogeneous Catalysts for the Selective Synthesis of Di- and Triglycerol 35
Arno Behr, Falk Lindner

Hydrogen Conversion by Reformation of Jet Fuel in a Supercritical Water Medium – Effects of the Water-to-Fuel Feed Ratio on the Conversion 36
Satya Putta, Jason W. Picou, Jonathan E. Wenzel, H. Bryan Lanterman, Sunggyu Lee

Needle in a Haystack Catalysis - An Experimental Study Using Temporal Analysis of Products (TAP) 43
Xiaolin Zheng, John T. Gleaves, Gregory S. Yablonsky, Thomas Brownscombe

Selectivity of Zirconia Supported Catalysts 50
Johnelle M. Halteman

Exceptionally Active Single-Site Mesoporous Pd/Al_2O_3 Catalysts for Allylic Alcohol Selective Aerobic Oxidation 51
Adam F. Lee, Rik M. Brydson, Simon F.J. Hackett, Jamil Khan, Peter Styring, Karen Wilson

Synthesis of Improved Fuel Cell Catalysts Using Electroless Deposition Methods 60
Kevin D. Beard, J.W. Van Zee, J.R. Monnier

Simulation of Heterogeneously Cs-Doped Heteropolyacid Catalyzed Transesterification for Biodiesel Production 67
Mian Xu, Karen Wilson, Jhuma Sadhukhan

Growth of Fe Based Nanocatalysts on Reconstructed Au(111) Surfaces 75
Christopher Matranga, Neetha Khan

Direct Dimethyether Synthesis Using Revese Shift Assisted Reforming 76
Kwang-Deog Jung, Oh-Shim Joo, Heon Jung, Dae Ryook Yang

Performance Comparison of a Bubble Reactor and a Loop Reactor for Preparing a C5 Hydrofluoroether, HFE-449mec-f, in the Gas-Liquid Biphasic Reaction 77
Hyunjoo Lee, Prabal Sapkota, Sang Deuk Lee, Hoon Sik Kim, Kwang Ho Song, Honggon Kim

Thermodynamics & Kinetics of the Erbium-Hydrogen System 78
Robert M. Ferrizz, Thomas J. Louck, M. Barry Ritchey, Loren I. Espada, Daniel R. Kammler, Clark S. Snow

Kinetic Model of Complex Reaction Network: Methanol Steam Reforming 80
Tai-Shang Chen, Jia-Ming Chern

Synthesis and Characterization of Au-Pt/SiO_2 Catalysts Prepared by Electroless Deposition Methods 81
Jayakiran Rebelli, Joseph H. Montoya, Melanie T. Schaal, Christopher T. Williams, John R. Monnier

In-Situ, Fiber-Optic Measurement Techniques in Carbon Dioxide Expanded Liquid (CXL) Multiphase Reactors 82
Sean G. Mueller, Muthanna H. Al-Dahhan, Milorad P. Dudukovic

Formation of Glycerol Derivatives by Catalytic Condensation and Acetalization 83
Aiguo Liu, Christopher Luckett, Hui Hui Li, Wayne Turbeville

Deep Oxidative Desulfurization of Fuels Catalyzed by Peroxophosphomolybdate Compounds in Ionic Liquids 84
Lining He, Huaming Li, Wenshuai Zhu, Junxiang Guo, Xue Jiang, Jidong Lu, Yongsheng Yan

Dramatic Micellar Catalaysis on the Chracteristics of B-Z Oscillatory Reaction with Citric Acid 88
Umesh Kumar Utkoor, Chinna Rajanna Kamatala

Oxidative Coupling of Methane on $BaTiO_3$ Perovskites Partially Substituted with Germanium 89
Marzieh Hamidzadeh, Davar Mohammadi Boghaei, Saeedeh Negin

Ultra-Deep Desulfurization of Fuels Catalyzed by Molybdic Compound in Ionic Liquids 95
Wenshuai Zhu, Huaming Li, Jiexiang Xia, Xue Jiang, Jidong Lu, Yongsheng Yan, Dan Xu

Comparison between Formic Acid Decomposition and CO Oxidation with Au Catalysts 99
Manuel Ojeda, Enrique Iglesia

Development of Nanoscale Ceria-Based Sulfur-Tolerant Catalyst Formulations 100
Desikan Sundararajan, Abdul-Majeed Azad

Flow Distribution in a High Pressure Trickle Bed Reactor 107
Zeljko V. Kuzeljevic, MIlorad P. Dudukovic, Muthanna H. Al-Dahhan

Activated Carbon Supported Catalysts and Adsorbents for Wgs Reaction 108
Burcu Selen-Çaglayan, A.Erhan Aksoylu

Design Implications of the Nonlinear Dynamics of the Monolithic Loop Reactor for Fischer-Tropsch Synthesis 109
Costin Sorin Bildea, Pietro Altimari, J. Ruud Van Ommen, Johan Grievink

Understanding Sulfur Poisoning in Steam Reforming of *n*-Hexadecane: Catalyst Characterization Studies 116
Satish Lakhapatri, Martin A. Abraham

Mathematical Modeling of Catalyst Deactivation by the Presence of Coke Formation in Catalyst Particles 117
M. Gaxiola-Cervantes, J. L. Rico-Cerda, M. Morales-Cabrera

Author Index

69g Enzymatic Synthesis of Small Chiral Alcohols in One-Phase and Two-Phase Systems

Pia Müller and Lasse Greiner. Institute of Technical and Macromolecular Chemistry, RWTH Aachen University, Worringer Weg 1, Aachen, 52074, Germany

Small enantiomerically pure alcohols are building blocks for pharmaceutical agents and important fine chemical intermediates. Therefore, they are of high commercial interest. However, high enantioselectivities are virtually impossible for the synthesis of small chiral alcohols with chemical catalysts.[1] Due to this, biocatalysis becomes more and more important for those applications. The enzymatic synthesis can be carried out in one-phase and two-phase systems with reaction conditions specially adapted to substrate and enzyme. In one-phase systems for the synthesis of small chiral alcohols the reverse reaction leads to decrease in enantiomeric excess (ee) and in conversion. Furthermore, in one-phase systems hardly water soluble substrates cannot be employed. Utilizing biphasic systems, equilibrium can be shifted to product side by extracting the products back into the non-aqueous phase[2], and racemisation can be supressed as well.[3] Apart from that, higher concentrations of substrates with low H2O solubility can be used (Fig. 1). The non-reactive phase also serves as a reservoir for the reactive phase. On the other hand, non-aqueous media can lead to deactivation of the enzyme. Results will be presented comparing one-phase and two-phase systems employing MTBE as a conventional organic solvent, and task specific ionic liquids (IL).[4] [1] Rissom, S.; Beliczey, J.; Giffels, G.; Kragl, U.; Wandrey, Chr. Tetrahedron: Asymmetry, 1999, 10, 923-928 [2] Eckstein, Marrit F.; Villela Vilho, Murillo; Liese, Andreas; Kragl, Udo Chemical Communication, 2004, 9,1084-1085 [3] Ansorge-Schumacher, Marion; Greiner, Lasse; Schroeper, Florian; Mirtschin, Sebastian; Hischer, Tanja Biotechnology Journal, 2006, 1, 564-568 [4] The kind gift of ionic liquids from Patrick Baeuerlein, University of Eindhoven, Netherlands, is acknowledged. The kind gift of enzymes from S. Na'amnieh and A. Pawlowski, X-Zyme GmbH, Duesseldorf, Germany, is acknowledged.

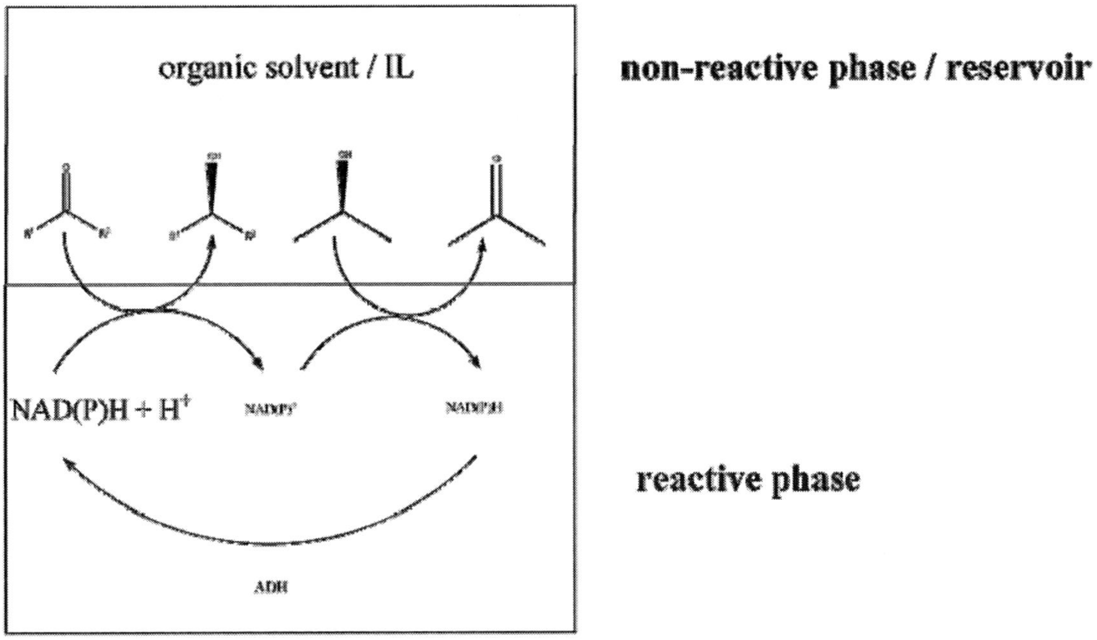

69c Hydroformylation of Higher Olefins with Immobilized Catalysts

Timo J. Koch, Jürgen Klankermayer, and Walter Leitner. institute of technical and marcomolecular chemistry, RWTH-Aachen University, Worringerweg 1, Aachen, Germany
In current industrial chemistry, aldehydes derived from higher olefins are important commodities for the synthesis of various fine chemicals. The resulting aldehydes from the hydroformylation processes are easily converted into many important secondary products like detergents, paints, coatings, plasticizers, copolymers, pharmaceuticals and agrochemicals.[1] The latest generation of industrial hydroformylation processes uses rhodium catalysts in combination with water soluble phosphines (Ruhrchemie/Rhône-Poulenc-process).[2, 3] A major drawback of this approach is its limitation to water soluble alkenes. A switch to other immobilization techniques often results in unacceptable high levels of catalyst leaching.[4] In this contribution, we describe that the combination of supercritical fluids with certain established immobilization techniques can open new process alternatives and reduce the amount of leaching significantly. In combination with supercritical fluids our immobilization approach leads to a flexible and stable reaction system for the hydroformylation of alkenes with various chain length (propene to octadecene).[5, 6] Moreover, the facile catalyst immobilization procedure enables the application of a wide range of transition metal complexes. [7, 8] The beneficial combination with supercritical fluids suppresses the catalyst leaching, allows the control of conversion and can be further developed to a continuous process.[9, 10] Literature: [1] B. Fell, Tenside, Surfactants, Detergents 1998, 35, 326. [2] L. Bexten, B. Cornils, D. Kupies, (Ruhrchemie A.-G., Fed. Rep. Ger.). Application: DE, 1986, p. 25 pp. [3] H. Bahrmann, B. Cornils, W. Konkol, W. Lipps, (Ruhrchemie A.-G., Fed. Rep. Ger.). Application: DE, 1985, p. 20 pp. [4] M. F. Sellin, I. Bach, J. M. Webster, F. Montilla, V. Rosa, T. Aviles, M. Poliakoff, D. J. Cole-Hamilton, Journal of the Chemical Society, Dalton Transactions 2002, 4569. [5] W. Leitner, Green Chemistry b Roger A. Sheldon, Vol. 407, 2000. [6] M. McCarthy, H. Stemmer, W. Leitner, Green Chemistry 2002, 4, 501. [7] B. Cornils, W. A. Herrmann, I. T. Horvath, W. Leitner, S. Mecking, H. Olivier-Bourbigou, D. Vogt, Editors, Multiphase Homogeneous Catalysis, Volume 2, 2005. [8] B. Cornils, W. A. Herrmann, I. T. Horvath, W. Leitner, S. Mecking, H. Olivier-Bourbigou, D. Vogt, Editors, Multiphase Homogeneous Catalysis, Volume 1, 2005. [9] N. Tanchoux, W. Leitner, Handbook of Green Chemistry and Technology 2002, 482. [10] P. g. Jessop, W. Leitner, Editors, Chemical Synthesis Using Supercritical Fluids, 1999.

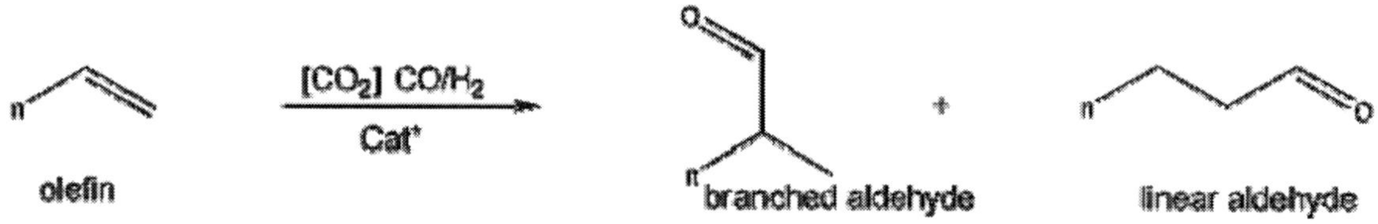

69b "Pot-in-Pot" Reactions: Polymer Macro-Fabrication for Site-Isolation of Catalysts and Reagents for Cascade Reactions

Martin T. Mwangi, M. Brett Runge, and Ned B. Bowden. Department of Chemistry, University of Iowa, 423I Chemistry Building, Iowa city, IA 52242

We reported the occlusion of Grubbs' catalysts in polymeric membranes leading to new reactivities and otherwise impossible Selectivities. Here we report the macro-fabrication and application of polymeric thimbles in "pot-in-pot" reactions. These newly discovered cascade reactions exploit selective diffusion of small molecules across a polymeric barrier to site-isolate organometallic catalysts, solvents, and/or reagents. This efficient site-isolation allowed us to perform new cascade reactions based on well understood reactions (see below). We have also applied our findings to perform well-controlled living polymerizations. Our findings will be discussed.

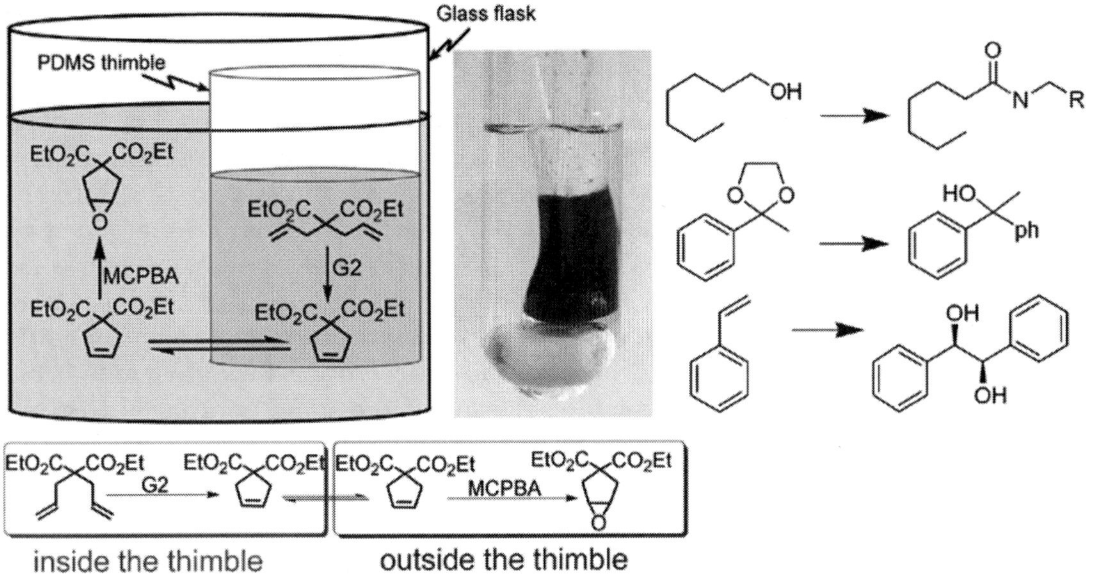

Reaction calorimetry as means for process intensification of the epoxidation methyl oleate

*Dipl.-Chem. Florian Pontzen, Dipl.-Ing. (FH) Daniela H. Müller, Prof. Dr. Marcel A. Liauw, Dr. Lasse Greiner**

Institut für Technische und Makromolekulare Chemie
RWTH-Aachen University, Germany

Abstract

Reaction calorimetry is a useful tool for process intensification since information about thermodynamics as well as kinetics of chemical reactions can be obtained simultaneously. The heat flow of a reaction gives insight into the properties and the behaviour of a reaction, especially when connected with a second analysis method. Heat flow calorimetry was successfully applied to the process intensification of the solvent-free epoxidation of methyl oleate with *in situ* generated peracetic acid. The productivity of the reaction could be increased 4 fold yielding the epoxide as a precursor for lubricants from renewable feedstock with comparable selectivity and yield compared to standard procedures.

Introduction

Reaction calorimetry

Reaction calorimetry is widely used for safety studies as well as the investigation of reaction kinetics and thermodynamics of chemical reactions and physical processes [1, 2]. Heat flow is directly proportional to the sum of reaction rates and reaction enthalpies

$$\dot{Q} = V \sum_i r_i \Delta H_i \qquad (1)$$

with the reaction rates r_i and reaction enthalpies H_i of all reactions i occuring in the reaction volume V, opposed to integral measurements for concentration or molar fractions,

*Corresponding author
greiner@itmc.rwth-aachen.de

respectively, obtained from GC, IR or other methods. The integrated heat flow signal, correlated with the reacted amount of substrate, corresponds to the molar reaction enthalpy, resp. the sum of all reaction enthalpies (including enthalpies of crystallisation, solution etc.)

$$\sum H_i = \frac{1}{x} \int_0^t \dot{Q} dt \tag{2}$$

By determination of the heat flow, which is proportional to the temperature difference of reactor and cooling jacket (*heat flow calorimetry*), or between inlet and outlet temperature of the cooling jacket (*heat balance calorimetry*), respectively, a direct measure for the reaction rate is given (differential signal). The problem of the lack of selectivity of the signal (all reactions are monitored simultaneously and cannot be separated) can be overcome by combination with a second (integral) analysis method. In our case, gas chromatography was chosen for chemical analysis of the reaction mixture.

Lubricants from biogenic feedstock

Fatty acid esters from biogenic feedstock can be used in a wide field of applications, including fuels (Biodiesel) nutrition and lubrication [3, 4]. In most cases a chemical modification has to be carried out due to limited stability of the oils. Besides ester hydrolysis, radicalic α-oxidation is the most important reaction leading to degradation of fatty acids and their derivatives [5].
A suitable method for the chemical modification of unsaturated fatty esters to yield biodegradable lubricants from biogenic feedstock is the epoxidation with peroxo compounds like peracetic or performic acid. By further derivatisation the properties of the lubricant can be altered leading to stable products for a wide range of applications. Epoxides can be easily converted by ring opening reactions with various nucleophiles yielding a wide range of products which are stable towards chemical and thermal attack [6, 7].

Results

Process intensification of the reaction in the calorimeter

The desired heat flow signal is obtained by correction of the raw data by a dosage compensation as well as reflux compensation including the temperature of the dosed substrate and the vapour temperature. A subsequent correction by including heat capacities and heat transfer coefficients of the reaction mixture as well as of the tempering medium (interpolated according to conversion between the values of starting and product mixture) yields the heat flow signal. In a typical calorimetric experiment a strong increase of the heat flow signal can be observed during dosage of H_2O_2, acetic and sulfuric acid (Figure 1). The maximum heat flow is about 61 W at 80°C. After that an

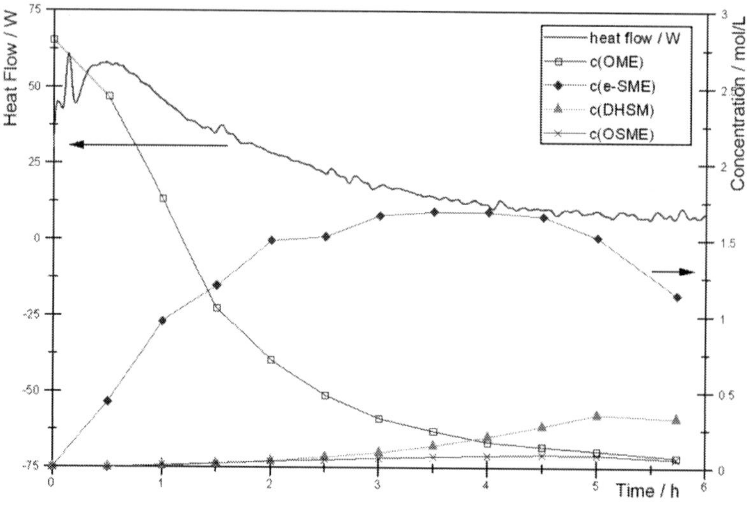

Figure 1: Heat flow curve of OME epoxidation, H_2O_2/OME = 2.5, 80°C, 400 rpm, 0.6 eq. acetic acid, 2 mol-% sulfuric acid

exponential decrease in heat flow signal can be observed. For these process conditions, the reaction is nearly complete after 4 hours giving about 93% conversion. The progress curve (integral of the heat flow signal) is obtained by baseline correction and integration.

The results obtained with *heat flow calorimetry* and *heat balance calorimetry* were similar. Since the heat flow signal obtained with the latter method showed a lower signal-to-noise-ratio, only the results received by *heat flow calorimetry* are shown.

Oleic acid methyl ester (OME, C18:1) (Figure 2) is epoxidised using peracetic acid, which is formed *in situ* in a biphasic system containing OME and a mixture of H_2O_2,

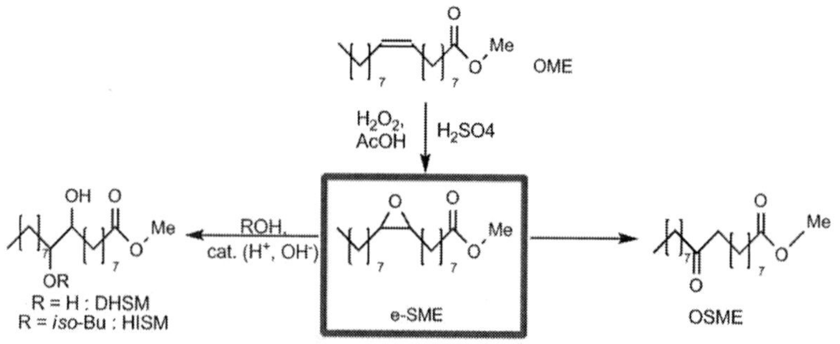

Figure 2: Reaction network of the OME system

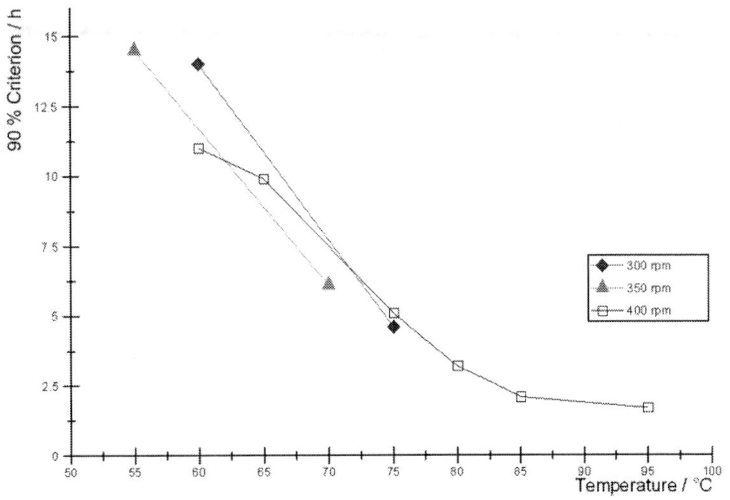

Figure 3: Temperature dependence of reaction time at different stirring speeds, H_2O_2/OME = 2.5, 0.6 eq. acetic acid, 2 mol-% sulfuric acid

acetic and sulfuric acid [8]. Besides the epoxidation many side and consecutive reactions occur, e.g. the successive hydrolysis to the dihydroxy-compound DHSM, rearrangement to a ketone OSME (Oxo-stearic acid methyl ester) as well as hydrolysis of the ester group.

According to a first optimisation the reaction was carried out at 60°C with a small excess of hydrogen peroxide and a reaction time of >10 h under stirring (400 rpm). Peracetic acid was formed *in situ* from H_2O_2 (35%), 0.6 eq. glacial acetic acid and 2% sulfuric acid [9]. In order to overcome the problem of too long reaction times, a systematic variation of temperature and stirring speed was done and the effect on productivity and selectivity was investigated (Table 1).

The results of the variations of temperature and stirring speed are shown in figure 3. It can be seen that the reaction is strongly dependent on the temperature. With increasing temperature the 90% criterion (time until 90% conversion is reached) decreases from 11 h at 60°C to 3.2 h at 80°C. The significant influence of the stirring speed giving hint to mass transfer limitation can only be observed at low temperatures. With increasing temperature the viscosity of the oil phase decreases which causes a negligible mixing sensitivity of the reaction at temperatures above 65°C.
Interestingly the selectivity towards e-SME does not change with increasing temperature, which was confirmed by GC analysis (Figure 1). Obviously the consecutive reactions (epoxide hydrolysis, rearrangement and ester hydrolysis) are not accelerated that much. The results obtained by GC and calorimetry match well. The correlation

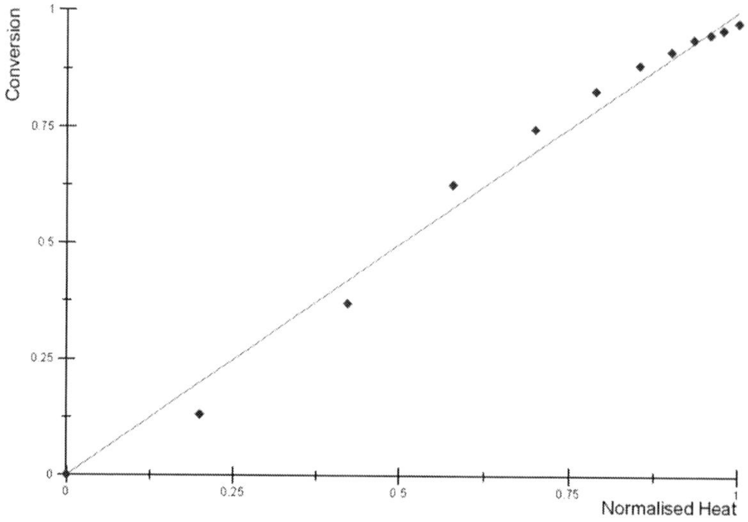

Figure 4: Correlation between predicted and measured conversion, 80°C, 400 rpm, H_2O_2/OME = 2.5, 0.6 eq. acetic acid, 2 mol-% sulfuric acid

between predicted conversion from the progress curve and the determined conversion from GC is nearly ideal (Figure 4). According to the results new optimised reaction conditions were set to 80°C at 400 rpm stirring speed with unchanged stoichiometry.

The overall average reaction enthalpy is $\Delta H = 235 \pm 27$ kJ mol^{-1} ($\pm$ 11%). Deviations are mainly caused by differences in H_2O_2 concentration due to decomposition during storage. The results match well with literature data for the epoxidation of linear Z-olefins, which were calculated with DFT methods [10]. The adiabatic temperature rise for the standard stoichiometry is 99 $\pm$ 19 K. The measurements yielded reasonable results which could be succesfully applied to process intensification.

Process intensification of the reaction in pilot scale

The improved reaction conditions could be successfully transferred to 25 L scale and lead to a 4 fold increase in productivity. A comparison between the reaction in the calorimeter and the pilot plant under similar conditions showed deviations of the heat flow signal in the pilot plant, mainly during the dosage due to inaccurate temperature control. Nevertheless, the reaction could be completed within 5 h and yielded products of nearly the same composition (Table 1).

Table 1: Product analysis from different syntheses

reactor	temperature/ °C	conversion/ %	selectivity/ %	reaction time/ h	STY/ mol/(L d)
calorimeter	60	95	86	18	1.2
calorimeter	80	92	86	4	5.4
pilot plant	60	95	86	18	1.0
pilot plant	80	94	85	5	3.6

Summary

It was shown that reaction calorimetry can be succesfully applied for process intensification studies for complex reaction systems and technical grade conversion. It allows safety assessments to achieve maximum operating safety of chemical processes. Calorimetry can be used to carry out an investigation and optimisation which is not accessible with other means. The direct observation of kinetics and thermodynamics of chemical reactions allows to obtain maximum information from a reaction which opens a wide range of opportunities, not only for industrial applicants. In our case the direct optimisation of a multiphase reaction was possible and lead to a 4 fold increase of productivity. The results could be transferred to 25 L scale .

Outlook

An overall aim of the work is to achieve a continuous operation in order to obtain a maximum productivity. For the continuous operation the application of microreaction technology is promising due to faster and more intense mixing as well as better heat transfer [11]. Furthermore the ecotoxicological and tribological properties of the products have to be investigated in order to develop a well suited biodegradable lubricant from biogenic feedstock [6, 12].

Acknowledgements

The heat balance calorimeter was provided by HiTec Zang (Herzogenrath, Germany). We kindly thank Prof. Dr. Werner Zang, Harald Beger, Carsten Joussen and Krzysztof Kucias for the technical support. Moreover we thank Marko J. S. Küppers for performing the reaction in the pilot plant. Financial support was provided by the graduate school

BioNoCo (GRK 1166) of the Deutsche Forschungsgemeinschaft (DFG) and the collaborative research center 442 (SFB 442).

Experimentals

1.4 mol OME (690 mL, Cognis, 70 mass-%) were charged, thermally equilibrated and calibrated to yield the heat capacities and the heat transfer coefficient. Then a mixture of 3.5 mol H_2O_2 solution (350 mL, Merck, Darmstadt, 35 mass-%), 0.87 mol glacial acetic acid (50 mL, Riedel-de-Haen, puriss. 99.8%) and 0.03 mol sulfuric acid (1.8 mL, Fluka, puriss. $\geq$ 95%) were dosed. For GC analysis, samples were taken and analysed after drying and dilution with dichloromethane/hexadecane. The measurements were carried out on a Restek Rtx®-Sil MS column (30 m, 0.28 mm ID) under hydrogen flow (150°C to 330°C). The product mixture was once again calibrated like described before.

In the pilot plant 23.6 mol OME (10 kg) were provided together with 14.2 mol glacial acetic acid (0.85 kg). After reaching isothermal conditions 59.0 mol H_2O_2 (5.73 kg, BASF, Ludwigshafen, 35 mass-%) together with 0.6 mol sulfuric acid (56.2 g) were added within 1.5 h.

References

[1] Heal, G. R.; Laye, P. G.; Price, D. M.; Warrington, S. B.; Willson, R. J.; *Principles of thermal analysis and calorimetry* **2000**, RSC press, Cambridge, Chapter *5*, 129-165.

[2] Zogg, A.; Stoessel, F.; Fischer, U.; Hungerbuehler, K.; *Thermochimica Acta* **2004**, *419*, 1-17.

[3] Mang, T.; Dresel, W. *Lubricants and Lubrication* **2001**, Wiley-VCH, Weinheim, Germany.

[4] Biermann, U.; Friedt, W.; Lang, S.; Lühs, W.; Machmüller, G.; Metzger, J.O.; Rüsch gen. Klaas, M.; Schäfer, H.J.; Schneider, M.P.; *Angew. Chem. Int. Ed.* **2000**, *39*(13), 2206-2224.

[5] Frankel, E. N.; *Autoxidation* in: *Fatty Acids* **1979**, edited by Everett H. Pryde , American Oil Chemists' Society, Champaign, Illenois, Chapter *18*, 353-378.

[6] Zhang, X.; Murrenhoff, H.; Weckes, P. P.; Hoelderich, W. *Synthetic Lubrication* **2004**, *21*(1), 3-11.

[7] Rios, L. A.; Weckes, P. P.; Schuster, H.; Hoelderich, W. F.; *Applied Catalysis* **2005**, *284*, 155-161.

[8] Greenspan, F. P.; Gall, R. J. *In situ epoxidation of organic esters with sulfuric acid and acetic acid*, USP 2,801,253 (**1953**).

[9] Eichholz, S.; *Doctoral Thesis: Prozessorientierte Untersuchungen zur Schmierstoffsynthese basierend auf Ölsäuremethylester und Glycerintrioleat* **2007**, RWTH Aachen University, Germany.

[10] Bach, R. D.; Dmitrenko, O.; Adam, W.; Schambony, S. *J. Am. Chem. Soc.* **2003**, *125*, 924-934.

[11] Jähnisch, K. Hessel, V.; Löwe, H.; Baerns, M. *Angew. Chem. Int. Ed.* **2004**, *43*(4), 406-446.

[12] Hahn, S.; Dott, W.; Eisenträger, A, *Journal of Synthetic Lubrication* **2006**, *23*(4), 223-236.

144e Rapid Lauroylation of Wheat Straw Hemicellulosesunder Microwave Heating

Junli Ren, South China University of Technology, Guangzhou, China, **Feng Xu**, Material Science and Technology, Beijing Forest University, Beijing, 100083, China, and Runcang Sun, Material Science and Technology, Beijing Forest University, Beijing, 100083, China.

Lauroylation of wheat straw hemicelluloses in the N,N-dimethylformamide/lithium chloride system under microwave irradiation was studied. The parameters optimized included lauroyl chloride concentration as the molar ratio of xylose unit in hemicelluloses/lauroyl chloride (1:1-1:4), 4-dimethylaminopyridine concentration (2-10%), reaction time (1-8 min), molar ratio of xylose unit in hemicelluloses/triethylamine (1:2), and reaction temperature (78°C). The reaction efficiency was measured by the yield and degree of substitution (DS). Under an optimum reaction condition (molar ratio of xylose unit in hemicelluloses/lauroyl chloride 1:3, molar ratio of xylose unit in hemicelluloses/ triethylamine 1:2, 5% 4-dimethylaminopyridine, 78°C, 5 min), a degree of substitution of 1.63 was obtained. The changes in the structure of hemicelluloses were verified by FT-IR, and 1H and 13C NMR spectroscopy. The result showed that the lauroylation occurred preferentially at the C-3 position of xylose unit in hemicelluloses. The behaviour of the lauroylated hemicelluloses was monitored by means of thermogravimetric (TG) and differential thermogravimetric (DTG) analysis. It was found that the product with low DS had a lower thermal stability than the native hemicelluloses, whereas the lauroylated polymers with high DS showed a higher thermal stability than the un-modified hemicelluloses. Keywords: Hemicelluloses; microwave; lauroylation; 13C NMR, thermal stability

205g Highly Stable Pt-BHA Nanocomposite Catalysts for Low-Temperature CO Oxidation

Yi Zhang[1], Sittichai Natesakawat[2], Tom Sanders[1], Christopher Matranga[2], and Goetz Veser[1]. (1) Department of Chemical Engineering, University of Pittsburgh, 1249 Benedum Engineering Hall, Pittsburgh, PA 15261, (2) National Energy Technology Laboratory of U. S. Department of Energy, P. O. Box 10940, Pittsburgh, PA 15236

Oxide-supported metal nanoparticles have attracted much attention in catalysis. In particular, strong size-effects in CO-oxidation over nano-sized gold catalysts have been extensively studied because of this reactions simplicity, fundamental character, and its importance in energy technology. However, most nanocatalysts to-date are plagued by insufficient stability at reactive conditions. To overcome this limitation, we are developing nanocomposites which combine the high reactivity of nano-sized noble metal particles with the stability of hexa-aluminates.

Here, we present results from an investigation in which Pt-BHA nanocomposites were synthesized and characterized (SEM, TEM, XRD, BET, and chemisorption), and tested in low-temperature CO oxidation. Kinetic studies via in-situ FTIR over a temperature range of T = 200 - 300 K show an exceptionally low activation energy of ~ 9 kJ/mol, about 3-times lower than for conventional Pt catalysts. CO-TPD studies confirm a strong size-effect for CO desorption energies in the range between 2 – 10 nm.

Kinetic modelling of *n*-dodecane dehydrogenation on Pt-Sn/Al_2O_3 catalyst based on LHHW mechanism

P. Valipour[1], M. Alavi[1], S. Rowshanzamir[2]
[1]Chemical Engineering Department, Iran University of Science and Technology (IUST).
[2]Green Research Center and Chemical Engineering Department, Iran University of Science and Technology, Tehran, Iran.

Abstract
Catalytic dehydrogenation of n-dodecane is explored experimentally and kinetics of the reactions over a promoted industrial *Pt–Sn/*Al_2O_3 catalyst between 460 and 490°*C* were determined. The main dehydrogenation reaction means to formation of mono olefins. Two other consecutive dehydrogenation reactions and also cracking of olefins to light paraffins took place. The rate models based on Langmuir-Hinshelwood-Hougen-Watson mechanism were derived for three dehydrogenation reaction and subjected to model discrimination and parameter estimation by using Genetic Algorithm optimization method. Of the 3 different models tested, model with surface reaction as rate-determining step fitted the experimental data well.

Keywords: Kinetics; Dehydrogenation; *n*-dodecane; LHHW; Mono olefin

---

1. Introduction
Dehydrogenation of higher paraffins (C_{10}-C_{14}) to the corresponding mono olefins on *Pt-Sn*/Alumina catalyst is an industrially important process and a basic step in the manufacture of bio-degradable detergents. Paraffin dehydrogenation is an endothermic reaction that is limited by chemical equilibrium. Higher conversion will require higher temperature and lower pressure [1]. The equilibrium conversion increases at higher temperatures, but side reactions, coke formation, and catalyst deactivation are also accelerated. The main reaction in catalytic dehydrogenation is the formation of mono-olefins from the corresponding feed paraffin [2].
The kinetic studies help to develop the reaction mechanism, establish reaction network, design of commercial reactors and evaluate catalyst efficiency and attempt improvements thereof. It became necessary to determine whether it was feasible to extend these relationships to description of the dehydrogenation kinetics for other paraffinic hydrocarbons in a commercial C_{10}-C_{14} fraction, which may be used as a feedstock in the production of higher linear olefins.
Very limited studies have been done on kinetics of higher paraffins dehydrogenation. Krylova et al [3-5] studied the mechanism and kinetics of dehydrogenation of *n*-decane on Platinum-Alumina catalyst promoted by Tungsten and Lithium. It seemed of interest to ascertain whether these laws are maintaibed when the chemical composition of the Platinum catalysts is varied. Sadykhova et al [6] investigated the kinetics and mechanism of the dehydrogenation of *n*-decane on a Platinum catalyst cantaining Tin promoter on a porous support. The laws governing the isotope exchange of Deuteriumin in the reaction components under the conditions of its realization were studied, the kinetic isotope effects were measured, the kinetic equations were found and the stepwise scheme

[1] Corresponding author. Tel/fax :(+9821) 77491242.
E-mail address: p_valipoor_80@yahoo.com

of the process was substatiated. Sadykhova [7] investigated the kinetics of dehydrogenation of *n*-undecane, *n*-dodecane, and mixtures of *n*-decane and *n*-dodecane on Platinum-Tin catalyst. It has been established that relative to mono metallic Pt / Al_2O_3 catalyst, bi metallic $Pt\text{-}Sn/Al_2O_3$ promotes desorption of olefins and hence may follow a different mechanism [7]. Gaidai et al. (1992) studied the dehydrogenation of n-decane in the nonsteady-state region and coking [8] and kinetics of coking during dehydrogenation of n-dodecane and n-decene–n-decane mixtures over a platinum-tin catalyst [9].
kinetic of coke formation of Basrur et al. (1998) [10] studied dehydrogenation of *n*-decane on promoted *Pt*-alumina catalyst using Box-Wilson experimental design and derived quadratic response surface correlations for predicting paraffin conversion and olefins selectivity for optimization of process conditions. Padmavthi et al (2005) [11] investigated the results of kinetics of *n*-dodecane dehydrogenation on a commercial catalyst based on promoted $Pt\text{-}Sn/Al_2O_3$ which were carried out in hydrogen gas recycle mode. Various rate models based on Langmuir-Hinshelwood-Hougen-Watson and Eley Rideal mechanisms have been derived for the three dehydrogenation reactions.
In this work, kinetic parameters of *n*-dodecane dehydrogenation for three consecutive dehydrogenation reactions based on Langmuir-Hinshelwood-Hougen-Watson investigated. The present study carried out in once through mode.

2. Experimental

2.1. Experimental set-up

The Experimental set-up is a fully automated system supplied by Vinci-Technologies, France. The schematic diagram of the experimental set-up used is shown in Figure 1. Major sections in the unit are: (1) Gas flow control modules (2) Liquid pumping system (3) Water pumping system (4) Pre-heater and mixer for hydrogen, *n*-dodecane and water (5) Reactor (6) Condenser and gas-liquid separator (7) Adequate flow meters to account material balance (8) GC analyzer. The reactor is an up-flow of homogenized hydrogen and *n*-dodecane vapor fixed bed reactor. The inner diameter of the reactor tube is 1.9 *cm*. Space above and below the catalyst in the reactor was filled with inert *SiC* (particle size 250 μm) to ensure proper distribution of fluid flow. The total bed height which is heated measures to 60 *cm*. The bed temperature indication is carried out by 4 thermocouples which were inserted in 4 heating zones. One thermocouple was also inserted in the center of the pre-heater.
The catalyst bed dilution with inert particles of the same size allowed uniform dispersion of the catalyst particles. Accurate mass flow controllers measured amounts of hydrogen in to and out of system. Feed to reactor and product from reactor are measured accurately up to second decimal by electronic balances.

2.2. Materials

The *n*-dodecane feed was supplied by Merck Schucharelt OHC, Germany and its purity is 99.65wt% and *n*-dodecene content of 0.258%. Arkan Gas Company supplies the hydrogen and nitrogen gas with purity of 99.999%. An industrial dehydrogenation catalyst is the product IFP company with the BET surface area of 116.7 m^2/g, pore volume of 0.504975 cm^3/g, and mean pore diameter of 13.8518 *nm*. The catalyst consists of highly porous spheoidal gamma alumina supprot containing 0.5 wt% of platinum and 0.7 wt% of Tin.

2.3. Experimental procedure

The reactor is first heated in a hydrogen flow of 10 Nl/hr at about 50 $^\circ C/hr$ to 430 $^\circ C$ to desorb the adsorbed moisture if any and maintain at this temperature for about 4 hours to enable reduction of platinum. The deionized water is pumped to the reactor before feed sent to the pre heater. Water injection through the catalyst bed is in the range of 1800-2000 *ppm* and promotes the conversion and selectivity of desired products. Fresh *n*-dodecane from storage tank is pumped by *HPLC* grade micro pump in to the pre heater. The feed is combined with fresh hydrogen from storage tank and water. The desired hydrogen to hydrocarbon ratio is maintained before the combined mixture is sent to the pre heater, which is maintained at $300^\circ C$. The pre-heater is designed to mix *n*-dodecane and water vapors and hydrogen homogeneously. The pre-heater outlet stream enters the reactor from bottom. The reactor outlet stream is cooled in a condenser and separated in to gas and liquid products in gas-liquid separator. The gas from the top of the gas-liquid separator is directed to the gas counter and GC analyzer. The condensed liquid product is collected at definite intervals of time in a storage tank.

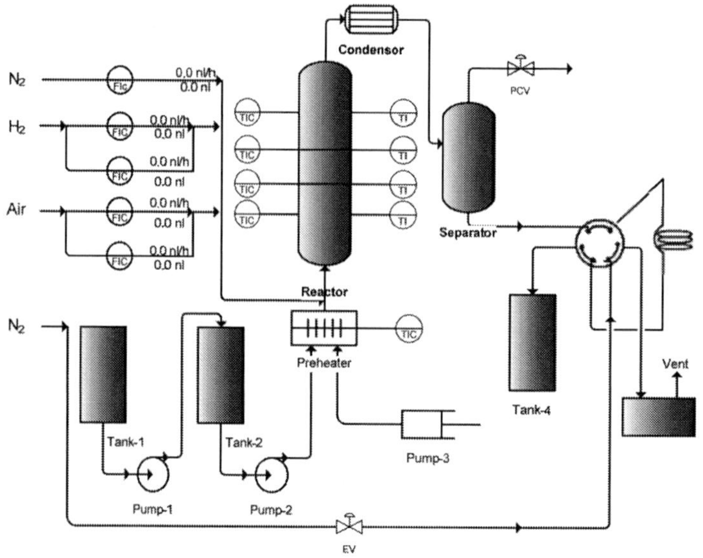

Figure 1. Schematic diagram of the experimental set-up

External mass transfer resistance can be minimizes by operating at a high velocity of gas and the pore diffusion resistance can be minimized by proper sizing of the catalyst. Absence of external mass transfer resistance was ensured by carrying out experiments at different mass flow rates and keeping *W/F* ratio constant. The conversion at two different liquid feed flow rate having same *W/F* ratio remains almost constant which indicates that the mass transfer effect on conversion in the given range of feed flow rate is negligible. The feed flow rate used in all experimental runs is in the range of 18-100 *ml/hr*. Preliminary runs were carried out to find out the particle size that allows neglecting internal transport limitation. Experiments were carried out by varying particle size and keeping *W/F* constant. The results did not show any significant deviation in conversion, which indicates the absence of intra particle diffusion.

Kinetic runs were carried out at different weight hourly space velocity, WHSV, H_2/HC ratio at three different temperatures of 460, 475 and 490 $^\circ C$. The WHSV is varied in the range of 20-111 (hr^{-1}). The H_2/HC ratio is varied in the range of 1-9. The gas products were analyzed using Agilent 6890N RGA GC and the liquid products were analyzed using Varian GC. The gases were analyzed using PLOT alumina column of 50 *m* length and 0.53 *mm* diameter capillary column and FID detector. Fused silica of 100 *m* length, 0.25 *mm* diameter and 0.5 μm film thickness and FID detector were used for liquid product analysis. The five groups identified in the order of column retention times are (1) C_5-C_{11} light ends from cracking reactions, (2) *n*-dodecane (reference peak), (3) mono olefins (five positional isomers are separated keeping end application in view), (4) conjugated di-olefins and non conjugated di-olefins are grouped as dienes (5) aromatics (primarily di-substituted types).

2.4. Experimental results

All the experiments were carried out in the absence of catalyst deactivation. The conversion and selectivity of various products are presented as a function of time in Figure 2. The results do not show any deactivation during the run time as can be observed from Figure 2. Each experiment is carried out with fresh catalyst to ensure same efficacy of the catalyst. To get the kinetic parameters corresponding to start of run conditions the first hour data was considered when the catalyst was fresh and free from deactivation. The material balance for all runs is in the range of 97.5-98.8%.

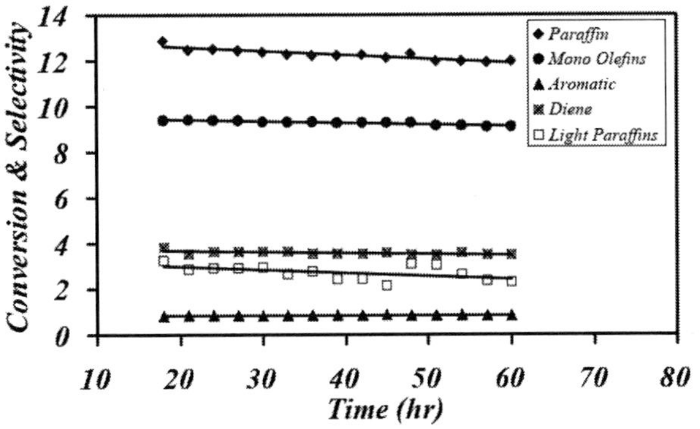

Figure 2. Effect of time on stream

3. Kinetic Analysis

3.1. Reaction scheme

Dodecene is the desired product in *n*-dodecane dehydrogenation. The liquid product of the reaction is about 99% and the rest is gas. All the products of n-dodecane dehydrogenation such as, dodecene, dodecadiene, etc. and *n*-dodecane itself can crack to form light paraffins. In this study, isomerization was ignored.

The reaction scheme corresponding to the three major dehydrogenation reactions have step for adsorption, surface reaction and desorption as each step could be the rate-determinig step. Therefore there could be three different rate equations.

Dehydrogenation
Olefin formation: $$C_{12}H_{26} \underset{k_{-1}}{\overset{k_1}{\leftrightarrow}} C_{12}H_{24} + H_2 \quad (1)$$

Secondary dehydrogenation

Diene formation: $$C_{12}H_{24} \underset{k_{-2}}{\overset{k_2}{\leftrightarrow}} C_{12}H_{22} + H_2 \quad (2)$$

Aromatic formation: $$C_{12}H_{22} \overset{k_3}{\rightarrow} C_{12}H_{18} + 2H_2 \quad (3)$$

Olefins cracking: $$C_{12}H_{24} + 2H_2 \overset{k_4}{\rightarrow} < C_{11} \quad hydrocarbons \quad (4)$$

The reaction scheme based on Langmuir-Hinshelwood-Hougen-Watson mechanism, shown in Table 1, was derived for three dehydrogenation reactions. For this mechansim three rate equations, shown in Table 2, were derived depending upon the assumed rate determining step.

Table 1. Reaction scheme for n-dodecane dehydrogenation

$C_{12}H_{26} + L \leftrightarrow C_{12}H_{26}L$
$C_{12}H_{26}L + L \leftrightarrow C_{12}H_{24}L + H_2L$
$C_{12}H_{24}L \leftrightarrow C_{12}H_{24} + L$
$C_{12}H_{24} + L \leftrightarrow C_{12}H_{24}L$
$C_{12}H_{24}L + L \leftrightarrow C_{12}H_{22}L + H_2L$
$C_{12}H_{22}L \leftrightarrow C_{12}H_{22} + L$
$C_{12}H_{22} + L \leftrightarrow C_{12}H_{22}L$
$C_{12}H_{22}L + L \leftrightarrow C_{12}H_{18}L + H_2 + H_2L$
$C_{12}H_{18}L \leftrightarrow C_{12}H_{18} + L$
$H_2L \leftrightarrow H_2 + L$

Table 2. Rate equations for LHHW model

$$r_1 = k_1\left(p_P - p_O p_H / K_{E1}\right)/DEN \qquad (adsorption \quad rds)$$
$$r_2 = k_2\left(p_O - p_D p_H / K_{E2}\right)/DEN$$
$$r_3 = k_3 p_D / DEN$$
$$DEN = \left(1 + K_P p_O p_H / K_{E1} + K_O p_D p_H / K_{E2} + K_D p_A p_H^2 + K_A p_A + p_H / K_H\right)$$

$$r_1 = k_1 K_P\left(p_P - p_O p_H / K_{E1}\right)/DEN \qquad (surface \quad reaction \quad rds)$$
$$r_2 = k_2 K_O\left(p_O - p_D p_H / K_{E2}\right)/DEN$$
$$r_3 = k_3 K_D p_D / DEN$$
$$DEN = \left(1 + K_P p_P + K_O p_O + K_D p_D + K_A p_A + K_H p_H\right)^2$$

$$r_1 = k_1 K_{E1}\left(p_P / p_H - p_O / K_{E1}\right)/DEN \qquad (desorption \quad rds)$$
$$r_2 = k_2 K_{E2}\left(p_P / p_H - p_D / K_{E2}\right)/DEN$$
$$r_3 = k_3 K_D p_D / DEN$$
$$DEN = \left(1 + K_P p_P + K_{E1} p_P / (K_O p_H) + K_{E2} p_O / (K_D p_H) + p_D / (K_A p_H^2) + p_H / K_H\right)$$

r_4 is same for all models and $K_{Ei} = k_i / k_{-i}$.

The procedure of establishing the mechanism of a chemical reaction are mainly based on comparison of experimental kinetic dependencies with theoretical equations, established on the basis of some model mechanisms and assumed to be valid under corresponding conditions. LHHW formalism was used for formulation of kinetics of slid catalyzed reversible dehydrogenation reaction. It was reported that through the assumptions made below may not always be fulfilled completely it is accepted that this approach is the most suitable and reliable way of rationalizing observed catalytic rate data [12]. The reaction scheme corresponding to the three major dehydrogenation reactions have steps for adsorption, surface reaction and desorption as shown in Table 1. The symbol L denotes the active site of catalyst surface. In step 1 molecular adsorption of *n*-dodecane takes place on catalyst active site. In the second step the adsorbed n-dodecane reacts with another active site to form $C_{12}H_{24}$-*L* and H_2-*L*. In third step dodecene is desorbed. The olefin is readsorbed (step 4) and diene is formed in stage 5 and 6. Dehydrogenation of diene takes place in a similar way. All reactions are assumed to have a step that is slower than the other steps, i.e. rate determining step. For example, if adsorption is considered as rate determining, then step 1, step 5 and step 9 in Table 1 are rate determining steps. The rate equations were derived I the same way similar to a single reaction. Rate equations were written for the rate determining sterp of each of the three reactions in terms of the concentrations of adsorbed species. These concentrations were then eliminated by means of the Langmuir equilibrium relations. This gives rate equations in terms of accessible gas-phase partial pressures and contains a denominator resulting from the adsorption of reacting species. Since all the three reactions are assumed to take place on the same active sites, the three rate equations have the same denominator.

4. Parameter Estimation

Since the reactor was operated in integral mode, parameter estimation was based on the minimization of the sum of residual absolutes of the conversion and selectivity of products by Genetic Algorithm optimization method. The calculated conversion and selectivity of products were obtained by numerical integration of the set of ordinary differential material balance equations of reference components.

4.1. Material balance

The rate of each component can be expressed as

Rate of disappearance of paraffin (*n*-dodecane):

$$\frac{dX_P}{d\theta} = r_1 \tag{5}$$

Rate of formation of olefins (dodecen):

$$\frac{dX_O}{d\theta} = r_1 - r_2 - r_4 \tag{6}$$

Rate of formation of diene (dodecadiene):

$$\frac{dX_D}{d\theta} = r_2 - r_3 \tag{7}$$

Rate of formation of aromatics:

$$\frac{dX_A}{d\theta} = r_3 - r_4 \tag{8}$$

Rate of formation of light paraffins ($<C_{11}$):

$$\frac{dX_{LP}}{d\theta} = r_4 \tag{9}$$

Rate of formation of hydrogen:

$$\frac{dX_{H2}}{d\theta} = r_1 + r_2 + 2r_3 - 2r_4 \tag{10}$$

Where $\theta = W/F =$ gcat/mol/h paraffin feed; r_i (i = 1-4) is rate of ith reaction. Reaction rate is expressed in terms of partial pressure of the components. The moles of different components present at any point when conversion of paraffin is X_P; moles of Paraffin 1-X_P; olefins = X_O; diene = X_D; aromatics = X_A; light paraffins = X_{LP}; hydrogen = X_{H2}. Partial pressure of each species can be expressed as

$$p_P = (1 - X_P)p_t/TM, \quad p_O = X_O p_t/TM, \quad p_D = X_D p_t/TM, \quad p_A = X_A p_t/TM, \quad p_{LP} = X_{LP} p_t/TM,$$

$$p_{H2} = X_{H2} p_t/TM, \quad TM = 7 - X_P + X_O + X_D + X_A + X_{LP} + X_{H_2}$$

Where TM is total moles in reaction mixture.

The above ordinary differential equations were integrated numerically by the forth order Runge-Kutta Method for a given set of kinetic prameters and operaring conditions to obtain concentration of each component at the end of reactor. The boundary conditions are

At $\theta = 0$; $X_i = 0$; $i = 1\text{-}n$ (n = no. of components)

Table 3. Results of objective function for three rate models

	Type of model	Rate determining step	Objective Function
1.	LHHW model (1)	Adsorption	23.46
2.	LHHW model (2)	Surface Reaction	7.40
3.	LHHW model (3)	Desorption	33.18

The kinetic parameters were estimated by minimization of the sum of the absoluts of the residuals of calculated and experimental values of conversion and selectivity of all the components by using Genetic Algorithm optimization method. The objective function is

$$\phi = \sum_{n}\sum_{m}\left|\frac{X_{n,m}calc - X_{n,m}\exp}{X_{n,m}\exp}\right| \tag{11}$$

When $X_{n,m}calc$ is the predicted selectivity of species m in the nth experiment and $X_{n,m}\exp$ is experimental selectivity value. The objective function values of all three rate models tested are presented in Table 3.

The conditions used to discriminate between the above three described models are: (1) if negative values were obtained for one or several parameters, the model was rejected, (2) objective function value should be the least of all the models, (3) when the objective function of two models are same, then the model with less number of parameters is considered.

Model discrimination is done by rejecting models based on their objective function, ϕ, values with minimum number of experimental data. LHHW rate model 3 based on desorption as rate determining step was rejected because of its high ϕ value. Both these retained LHHW models with adsorption and surface reaction as rate determining step were considered for further processing with complete data from further experimentation that allows to check the parametric selectivity of the model. The final objective function value of the rate model with surface reaction as rate determining step obtained from Eq. (11) is 27 as compared to 54, for rate model with adsorption as rate determining step, thereby eliminating the LHHW model with adsorption as rate determining step.

The rate of reaction was calculated from the W/F vs. conversion data. The conversion of each reaction was calculated by material balance. The data were fitted in to a smooth logarithmic curve of the type $X = m * \ln(W/F) + C$. Then the equation of differentiated $dX/d(W/F)$ to obtain rate as a function of W/F. The rate data indicate that the rate of formation of aromatics is proportional to concentration of dienes and does not depend on the concentration of olefins and paraffins which also supports the stepwise conversion in order paraffin-olefin-diene-aromatics. Figure 3 presents plots of the dependence of the reaction rate on the overall extent of conversion. The rate decreased linearly with conversion at the three temperatures. The curve have a somewhat concave shape, which may indicate inhibition of the reaction by any of its products, i.e. by olefins, dienes or aromatics [13].. The results show that the desorption of the products may not be controlling the rate in the present case which was confirmed from objective function values.

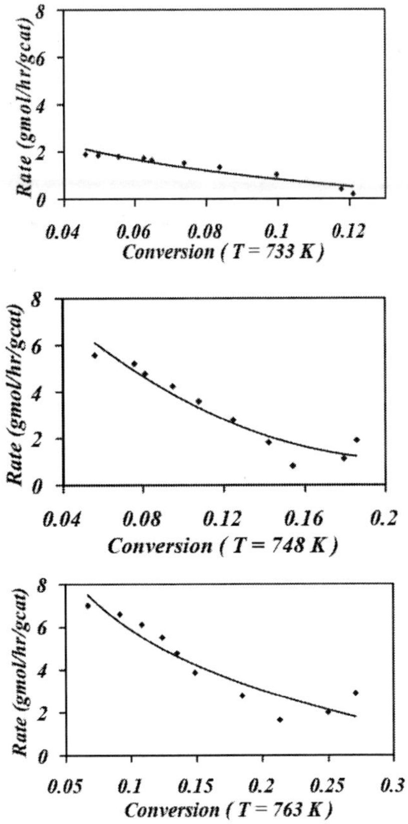

Figure 3. Rate vs. conversion of paraffin

Based on the observation drawn from rate data and analysis of objetive function values as discussed above, rate model with surface reaction as rate determining step was retained as final model for the process. The rate expression according to the retained model is

$$r_1 = k_1 K_P \left(p_P - p_O p_H / K_{E1}\right) / \left(1 + K_P p_P + K_O p_O + K_D p_D + K_A p_A + K_H p_H\right)^2 \tag{12}$$

$$r_2 = k_2 K_O \left(p_O - p_D p_H / K_{E2}\right) / \left(1 + K_P p_P + K_O p_O + K_D p_D + K_A p_A + K_H p_H\right)^2 \tag{13}$$

$$r_3 = k_3 K_D p_D / \left(1 + K_P p_P + K_O p_O + K_D p_D + K_A p_A + K_H p_H\right)^2 \tag{14}$$

$$r_4 = k_4 p_O \tag{15}$$

Where K_P, K_O, K_D, K_A and K_H are adsoption equilibrium constants and are related to equilibrium constants of the steps of the reaction in the following way:

$$K_P = K_1 \,,\quad K_O = 1/K_3 \,,\quad K_D = 1/K_6 \,,\quad K_A = 1/K_9 \,,\quad K_H = 1/K_{10}$$

The Arhenius plot of the rate constants is given in Figure 4. Adsorption equilibrium constants decreased with increase in temperature as it should. Hence the estimates of activation energies have to be statistically positive and those for adsorption enthalpies negative [12]. The plot of adsorption constants vs. $1/T$ is given in Figure 5. The activation energy and pre-exponential factor values of the retained model for all the reactions are in the same range with that reported for n-decane dehydrogenation [4, 5].

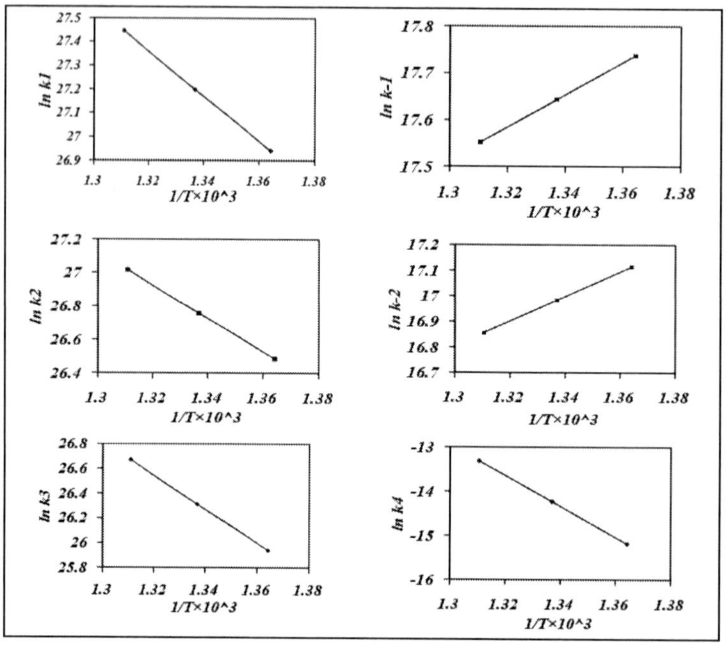

Figure 4. Arhenius plots of kinetic parameters

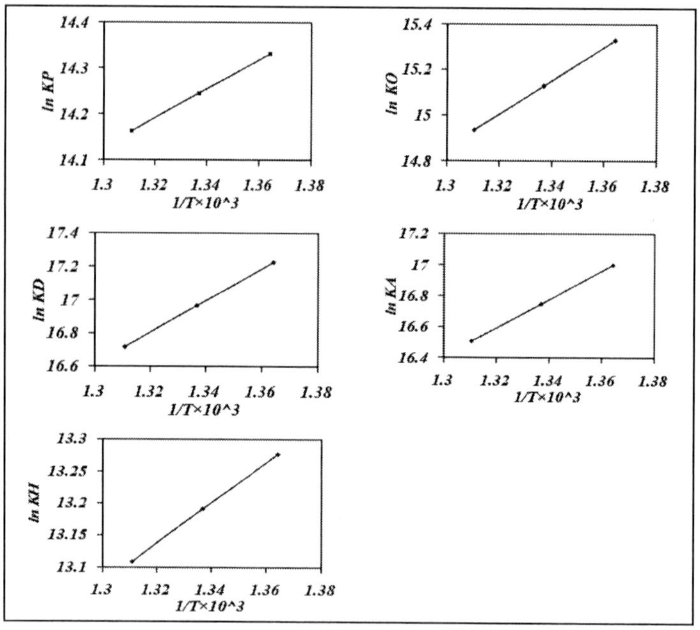

Figure 5. Adsorption constants vs. temperature

The values of the kinetic parameters are given below:

$k_1 = 1.86\times10^{17} \exp(-78.13 / RT)$, $mol\ Pa^2/h/g\ cat$

$k_{-1} = 4.56\times10^{5} \exp(28.69 / RT)$, $mol\ Pa/h/g\ cat$

$k_2 = 2.38\times10^{17} \exp(-82.44 / RT)$, $mol\ Pa^2/h/g\ cat$

$k_{-2} = 3.78\times10^{4} \exp(40.07 / RT)$, $mol\ Pa/h/g\ cat$

$k_3 = 2.18\times10^{19} \exp(-113.27 / RT)$, $mol\ Pa^2 / h / g\ cat$

$k_4 = 1.22\times10^{14} \exp(-290.24 / RT)$, $mol\ Pa^{-1} / h / g\ cat$

$K_P = 2.26\times10^{4} \exp(26.26 / RT)$, Pa

$K_O = 1.99\times10^{2} \exp(61.14 / RT)$, Pa

$K_D = 7.35\times10^{1} \exp(78.78 / RT)$, Pa

$K_A = 8.61\times10^{1} \exp(76.44 / RT)$, Pa

$K_H = 1.76\times10^{2} \exp(26.07 / RT)$, Pa

5. Effect of operating parameters

Figure 6 and 7 shows the effect of *W/F* on conversion of paraffin and selectivity of mono olefins at three different temperatures. At a constant *W/F* paraffin conversion increased with increase in reaction temperature. In the paraffin dehydrogenation reaction, the selectivity of olefin formation decreases steadily with increasing conversion. The selectivity of mono and di-olefins decreased with increase in temperature. The decrease in selectivity of mono olefins with increase in conversion, supports the stepwise mechansim of the paraffin dehydrogenation reaction [4-6]. The loss of olefins selectivity is due to secondary dehydrogenation to diene and then to aromatic hydrocarbons and parallel formation of light paraffins by olefin cracking and finally to coke. The selectivety of aromatics and light paraffins increased with increase in temperature.

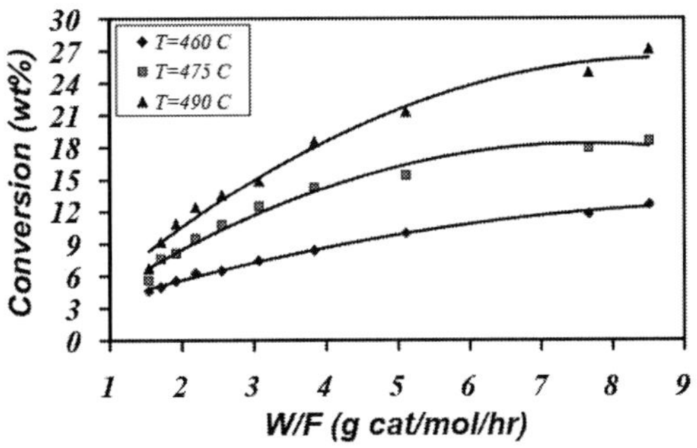

Figure 6. Effect of W/F on Conversion

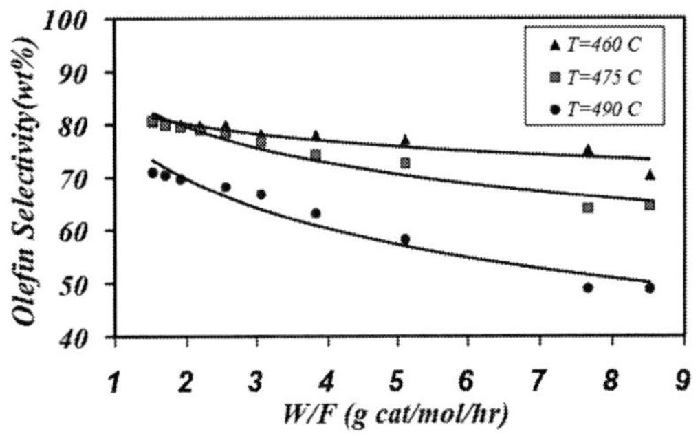

Figure 7. Effect of W/F on Selectivity

Figure 8 presents plots of the dependence of the changes in the concentration of the reaction products on the value of the space velocity. For comparison, the figure also presents analogous curves for the reaction on the platinum-alumina [3-5] and platinum-tin/alumina catalyst [6, 7]. The nature of curves corresponding to the accumulation of each of the products is identical. However, the rate of the formation of the olefins on the platinum-tin catalyst is higher, and the rate of the accumulation of the aromatic hydrocarbons is significantly lower. The effect of H_2/HC ratio on paraffin conversion and selectivity of the products is shown in Figure 9. The paraffin conversion decreased with increase in H_2/HC ratio. The decreased H_2/HC ratio at a constant pressure lowers the hydrogen partial pressure, which favors higher conversion. It can be seen that with decreasing concentrationof hydrogen in the reaction mixture, the rate of formation of aromatics increases to the greatest degree, and the rate of formation of olefins increases to the smallest degree [7]. This can be happen only through an increase in the concentration of the predecessor products, i.e. with a consecutive scheme.

Thus, the kinetic data obtained provide evidence of the consecutive nature of the process and of the absence of its inhibition by aromatic hydrocarbons. Since an increase in the concentration of hydrogen causes a reversible decrease in the reaction rate, the concentration of hydrogen appears in the denominator of the kinetic equations. In addition, the concentration of at least one of the other reaction products should appear in the denominator of the kinetic equations, since the conversion curves obtained at a constant hydrogen concentration have a concave form.

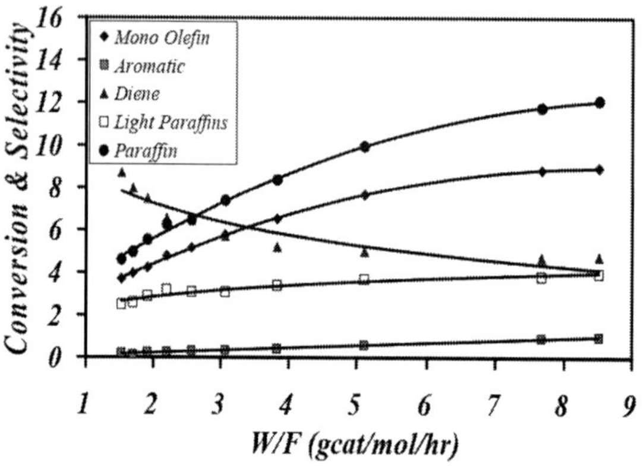

Figure 8. Dependence of the content of the products of the dehydrogenation of n-C_{12} vs. space velocity at 460 $^{\circ}C$

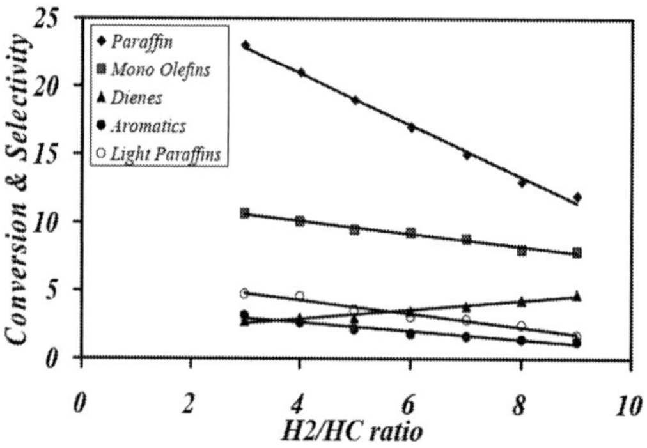

Figure 9. Effect of H2/HC on Conversion and Selectivity

6. Conclusion

Kinetics of n-dodecane dehydrogenation on promoted industrial $Pt\text{-}Sn/Al_2O_3$ catalyst was investigated. The results support the stepwise mechanism of the reaction. Three sets of rate models based on LHHW mechanism for all the three dehydrogenation reactions were derived and subjected to parameter estimation and model discrimination. The kinetic parameters were estimated and determined using Genetic algorithm optimization method. The kinetic parameters of the selected model are statistically significant and follow thermodynamic laws.

7. Acknowledgments

This work was supported by Iran University of Science and Technology (IUST) and Iran Polymer and Petrochemical Institute (Petrochemical Department).

8. References

1. M.M. Bhasin, J.H. McCain, B.V. Vora, T. Imai and P.R. Pujadó, "Dehydrogenation and oxydehydrogenation of paraffins to olefins", *Applied Catalysts A*: General 221 (2001) 397-419.
2. Domenico Sanfilippo , Ivano Miracca, "Dehydrogenation of paraffins: synergies between catalyst design and reactor engineering", *Catalysis Today 111 (2006) 133–139.*
3. T.L. Krylova, N.V. Nekrasov, B.S. Gudkov, N.A. Gaidai, V.R. Gurevich and S.L. Kiperman, "Mechanism of the Dehydrogenation of n-decane on Aluminum-Platinum Catalysts", in: Supported Metallic Catalyst for the Conversion of Hydrocarbons, Inst. Kataliza SO AN SSSR, Novosibirsk, pp. 307-310, 1978.
4. T.L. Krylova, N.V. Nekrasov, B.S. Gudkov, V.R. Gurevich and S.L. Kiperman, "Mechanism and Kinetics of Dehydrogenation of Higher n-Paraffins on Promoted Platinum Catalysts. I. Deuterium-Hydrogen Exchange of n-decane and Kinetic Isotope Effects in its Dehydrogenation Reaction on an Aluminum-Platinum Catalyst", (Translated from *Kinetika I Kataliz*, Vol. 21, No. 6, pp. 1482-1486, 1980).
5. T.L. Krylova, N.V. Nekrasov, N.A. Gaidai, V.R. Gurevich and S.L. Kiperman, "Mechanism and Kinetics of Dehydrogenation of Higher n-Paraffins on Promoted Platinum Catalysts. II. Study of Kinetics of the Dehydrogenation of n-decane on a Aluminum-Platinum Catalyst", (Translated from *Kinetika I Kataliz*, Vol. 21, No. 6, pp. 1487-1493, 1980).
6. Z.A. Sadykhova, N.V. Nekrasov, V.R. Gurevich and S.L. Kiperman, "Mechanism and Kinetics of Dehydrogenation of Higher n-Paraffins on Promoted Platinum Catalysts. III. Study of Kinetics and Mechanism of the Dehydrogenation of n-decane on Platinum-Tin Catalyst", (Translated from *Kinetika I Kataliz*, Vol. 22, No. 2, pp. 396-402, 1981).
7. Z.A. Sadykhova, N.V. Nekrasov, V.R. Gurevich and S.L. Kiperman, "Mechanism and Kinetics of Dehydrogenation of Higher n-Paraffins on Promoted Platinum Catalysts. V. Kinetics of Dehydrogenation of n-Undecane and n-dodecane on Platinum-Tin Catalyst", Vol. 25, No.3, pp. 593-597, 1984.
8. N.A. Gaidai, S.K. Mart'yanova ans S.L. Kiperman, "Mechanism and Kinetics of Dehydrogenation of Higher n-Paraffins on Promoted Platinum Catalysts. VI. Dehydrogenation of n-decane in the nonSteady-State Region and Coking over a Platinum-Tin Catalyst" (Translated from *Kinetika I Kataliz*, Vol. 33, No. 1, pp. 230-235, Jan.-Feb. 1992).
9. N.A. Gaidai, S.K. Mart'yanova ans S.L. Kiperman, "Mechanism and Kinetics of Dehydrogenation of Higher n-Paraffins on Promoted Platinum Catalysts. VII. Kinetics of Coking during Dehydrogenation of n-dodecane and n-decene–n-decane Mixtures over a Platinum-Tin Catalyst" (Translated from *Kinetika I Kataliz*, Vol. 33, No. 1, pp. 235-238, Jan.-Feb. 1992).
10. A.G. Basrur, D. Rajeshwer, D.T. Gokak, G.S. Rao and K.R. Keishna Murthy, "Recent advances in basic and applied aspects of industrial catalysis. Modelling of n-decane Dehydrogenation Using Box-Wilson Experimental Design", Vol. 113, pp. 809-814, 1998.
11. G. Padmavathi, K.K. Chaudhuri, D. Rajeshwer, G. Sreenivasa Rao, K.R. Krishnamurthy, PC. Trivedi, K.K. Hathi and N. Subramanyam, "Kinetics of *n*-dodecane dehydrogenation on promoted platinum catalyst", *Chemical Engineering Science, 60 (2005) 4119 – 4129.*
12. G.F. Froment and L.H. Hosten, "Catalysis Science and Technology", in: Catalytic Kinetics Modelling, Springer, Berlin, Heidelberg, New York, pp. 97-170 (Chapter 3), 1981.
13. S.G. Bashkirova and S.L. Kiperman, "Analysis of Kinetic Equations", *Kinetika I Kataliz*, Vol. 11, No. 3, 631-637, 1970.

Aerobic Epoxidation of Alkenes with Manganese (III) Salen Complexes Immobilised Onto Aminopropyl-HMS

Jamil Khan[1], Peter Styring[1] and Adam F. Lee[2],
(1)Department of Chemical and Process Engineering, University of Sheffield, Hadfield Building, Mappin Street, Sheffield, United Kingdom,
(2)Department of Chemistry, University of York, Heslington, York, United Kingdom

ABSTRACT

A multi-grafting technique has been used for the anchoring of new salen manganese complexes, Mn-Sal-Lys onto modified hexagonal mesoporous silica (HMS) with aminopropyl groups and 4-(aminomethyl)benzoic acid. The HMS material was analysed with FT-IR, N_2 adsorption and XPS. Along with two homogeneous complexes, the catalytic activity was tested for the aerobic epoxidation of cyclohexene and styrene to their oxides using isobutyraldehyde as a co reductant under mild conditions. The homogeneous catalysts were more active however it is possible to get good yields and selectivity from the immobilised catalysts. Repeated experiments were conducted and even after the fifth run, the results show some consistency with some drop in performance.

1. INTRODUCTION

Selective oxidation and epoxidation reactions have been of interest to organic synthetic chemists over the last sixty years. With a wide range of unsaturated compounds available, early work concentrated on the epoxidation of fatty materials with perbenzoic acid [1] and peracetic acid [2] to form epoxystearic acid. There was considerable research into this field for many decades with the focus shifting to the use of biological systems of animal and insect microsomes for the epoxidation of insecticides [3-5] and vitamins [6] in the mid-60s. In the late 70s and early 80s Groves synthesised an iron porphyrin and catalysed the asymmetric epoxidation of styrene using iodosylbenzene as the oxidant [7-9]. However porphyrins are difficult to make and offer low enantioselectivity, hence salen-type metal complex catalysts were investigated for their use in epoxidation reactions. Even though many metals complex with salen ligands only a few, manganese(III) [10-12], nickel(II) [13], chromium [14] have been used for epoxidation reactions. The main problems with these systems are the use of homogeneous catalyst which makes product separation difficult, and the oxygen source which tends to be a peroxy acid.

Mesoporous inorganic materials have been extensively used for heterogeneous catalysis and adsorption media with pores in the range of 20–500 Å due to their large internal surface area [15]. Tanev and Pinnavaia synthesised a variety of silicates using ionic and neutral pathways and compared the properties by a variety of physical techniques [16]. The neutral pathway (S°O°) was achieved at ambient temperatures due to the lower assembly forces and afforded hexagonal mesoporous silicates (HMS). The main advantages of the HMS method are their greater thermal stability and thicker walls. The HMS material was hailed as having the potential of superior adsorption and catalysis by guest molecules.

Various techniques have been used to anchor the metal complex onto a solid support, Kim and Kim immobilised Mn complexes onto MCM-41 *via* an ion exchange method and this was used for the epoxidation of styrene with *m*-chloroperoxobenzoic acid (*m*-CPBA) as the terminal oxidant [17]. Benedito *et al.* managed to immobilise metalloporphyrins onto functionalised 4Å molecular sieves [18]. The surface of the silica was functionalised with aminopropyl groups and the porphyrin was attached to the surface *via* a peptide-type linkage through the carboxylate group on the porphyrin to the aminopropyl-modified molecular sieve. Silva *et al.* immobilised *Jacobsen*-type salen complexes onto HMS using two methods; (i) one was anchored directly through the carboxylic acid functionalised diimine bridge fragment of the complex and (ii) the other through the hydroxyl group on the salicylaldehyde fragment of the complex, initiated by cyanuric chloride [19]. Zhang and Li Also found that the linker length can affect the results of asymmetric epoxidation with the enantioselectivity raising from 44.9% ee to 82.6% ee by increasing the bond length from 1 to 5 units [20].

The purpose of this work is to investigate the potential of novel heterogeneous catalysts for the epoxidation of olefins. This study will investigate the immobilisation of salen complexes onto well defined mesoporous silicates and subsequent tests for the epoxidation of olefins. This study will also try to employ reaction conditions which are readily available using mild conditions and materials with the greatest accessibility. In this report the catalyst was immobilised using a multi-grafted technique. The epoxidation of the olefins was conducted using the method developed by Yamada and his co-workers [21]. We also looked at the effect of increasing the linker length from aminopropyl-modified HMS to phenyl- and cyclohexyl-modified HMS.

2. EXPERIMENTAL

2.1. *Materials*

Tetraethyl orthosilicate (98%), dodecylamine (98%), 3-aminopropyl triethoxysilane, 4-(aminomethyl)benzoic acid (97%), 3,5-Di-*tert*-butyl-2-hydroxybenzaldehyde (99%), DL-Lysine (≥98%), Cyclohexene (99 %), Styrene (99%), (1R,2R)-1,2-cyclohexanediamine, manganese acetate tetrahydrate ($Mn(OAc)_2.4H_2O$, (99%), *N*-Hydroxyphthalimide 97% and *N,N'*-Diisopropylcarbodiimide 99% were purchased from Sigma-Aldrich. Lithium chloride, sodium hydroxide, ethanol absolute, dichloromethane, diethyl ether and acetonitrile were purchased from Fisher Scientific Ltd. All chemicals were of reagent grade and used as received without further purification.

2.2. *Characterisation and Analysis*

The gas chromatography (GC) was performed using a Varian 3900 GC with a 15 m × 25 mm i.d. column. For the work with styrene a ramp rate of 10 °C/min from 60 °C to 150 °C and for cyclohexene a ramp rate of 10 °C/min from 30 °C to 100 °C were used. Thermal Gravimetric Analysis (TGA) was conducted with a Stanton Redcroft STA-780 Series Thermal Analyser over a temperature range of 20 °C to 700 °C at 20 °C/min under a 20 mL/min He flow. The pore size distribution and specific surface area was analysed using Micrometrics ASAP 2010 instrument using the BET isotherm with N_2 as the adsorbent. CHN analysis was conducted with Perkin Elmer 2400 CHNS/0 Series II Elemental Analyzer. FT-IR was conducted with Perkin Elmer Spectrum One Fourier Transform Infra-red (FTIR) spectrometer in the region of 450-4000 cm^{-1}. The samples were diluted with KBr before analysis.

2.3. *Synthesis and modification of hexagonal mesoporous silica*

2.3.1. *Synthesis of hexagonal mesoporous silica*

The hexagonal mesoporous silica (HMS) was synthesised as reported in literature [16],. Tetraethoxysilane (TEOS, 18.75 g, 0.09 mol) was dissolved in a solvent mixture of ethanol (41 g) and distilled water (53 g) into a 500 mL round-bottomed flask. To the mixture, n-dodecylamine (5.09 g) was added turning the solution milky and this was stirred for 18h at ambient temperature. The resulting white suspension was filtered and refluxed in ethanol (100 mL) for 3 h to extract the n-dodecylamine template. The template removal was repeated three times. The reaction yielded approximately 10 g of HMS.

2.3.2. *Functionalisation of hexagonal mesoporous silica*

The functionalisation of the HMS with 3-aminopropyltriethoxysilane was adapted from Silva *et al* [19]. 3 mmol of 3-aminopropyltriethoxysilane was added to 1 g of the solvent extracted HMS in dry toluene and refluxed for 24 h. The resulting material (AMPS-HMS) was vacuum-filtered and washed several times with 50 mL toluene and dried overnight in an oven at 110 °C.

2.3.3. *Immobilisation of phenyl-groups onto AMPS-HMS*

1.50 g of AMPS-HMS in 20 mL 1:1 DCM/DMF was stirred with 0.45 mmol 4-(aminomethyl)benzoic acid with 0.22 mmol *N*-hydroxyphthalimide. After 40 minutes at room temperature, 0.80 mmol *N,N'*-Diisopropylcarbodiimide was added. The reaction was continued for a period of 24 h. The respective silica was separated using vacuum-filitration and washed once with 10 mL DMF and twice with 10 mL DCM and dried overnight in an oven at 60 °C. The 4-(aminomethyl)benzoic acid modified HMS will be referred to as 4MBA-HMS.

2.4. *Synthesis of the manganese(III) salen complexes*

2.4.1. *Jacobsen complex*

The complex used for the preparation of Jacobsen's catalyst was adopted from literature [22]. A 250 mL three-necked round-bottomed flask equipped with a magnetic stirrer and reflux condenser was charged with 4.0 mmol 3,5-Di-*tert*-butyl-2-hydroxybenzaldehyde in 20 mL 95% ethanol. To this, 2.0 mmol (1R,2R)-1,2-cyclohexanediamine in 10 mmol 95 % ethanol was added to the flask and refluxed for of 95% EtOH was added. The flask was refluxed for 1 h resulting in the formation of a bright yellow precipitate. A solution of 0.5 M KOH (8.0 mL, 4.0 mmol) in 95% EtOH was added dropwise. Once all of the contents had dissolved a solution of $Mn(OAc)_2.4(H_2O)$, (5.0mmol) in 5.0 mL water was added. The reaction mixture turned brown,was continued at reflux for 30 min. The reaction was allowed to cool for another at which 3 equiv. of LiCL was added in 5 mL of water and left for an hour. The reaction mixture was vacuum filtered through a Buchner funnel with a fritted disk, porosity 4. The filtrate was reduced in volume to a third using a rotary evaporator and the complex was precipitated from solution by adding it to 50 mL of water in a 100 mL Erlenmeyer flask. The precipitate was separated using filtration and recrystallised with hot acetonitrile.

2.4.2. *Mn-Sal-Lys complex*

The Mn-Sal-Lys complex was developed from reports in literature [23, 24]. 1.0 mmol of DL-Lysine and 1.0mmol of KOH were stirred in 10 ml of 95% EtOH and refluxed. Upon dissolution, 2.0 mmol of 3,5-Di-*tert*-butyl-2-hydroxybenzaldehyde in 20 ml of the EtOH was added. The solution became yellow and after stirring for 1 h, 10 ml of the solution of 1.0 mmol $Mn(OAc)_2.4H_2O$ was slowly added. The colour of solution changed to brown and the solution was continuously stirred for another hour. Three equivalents of LiCL were added and the reaction mixture was allowed to cool to room temperature. The solvent was removed using a rotary evaporator and the products recrystallised from acetonitrile.

2.5. Immobilisation of Mn-Sal-Lys onto modified-HMS

The following is typical for the covalent attachment of Mn-Sal-Lys onto either AMPS-HMS and 4MBA-HMS summarised in Scheme1. To 0.5 g of AMPS-HMS in 10 ml of 1:1 DCM/DMF, was stirred with 0.15 mmol Mn-Sal-Lys with 0.073 mmol *N*-hydroxyphthalimide. After 40 minutes at room temperature, 0.323 mmol *N,N'*-Diisopropylcarbodiimide was added. The reaction was continued for a period of 24 h. The immobilised silica was separated using vacuum-filitration and washed once with 10 mL DMF and twice with 10 mL DCM and dried overnight in an oven at 60 °C.

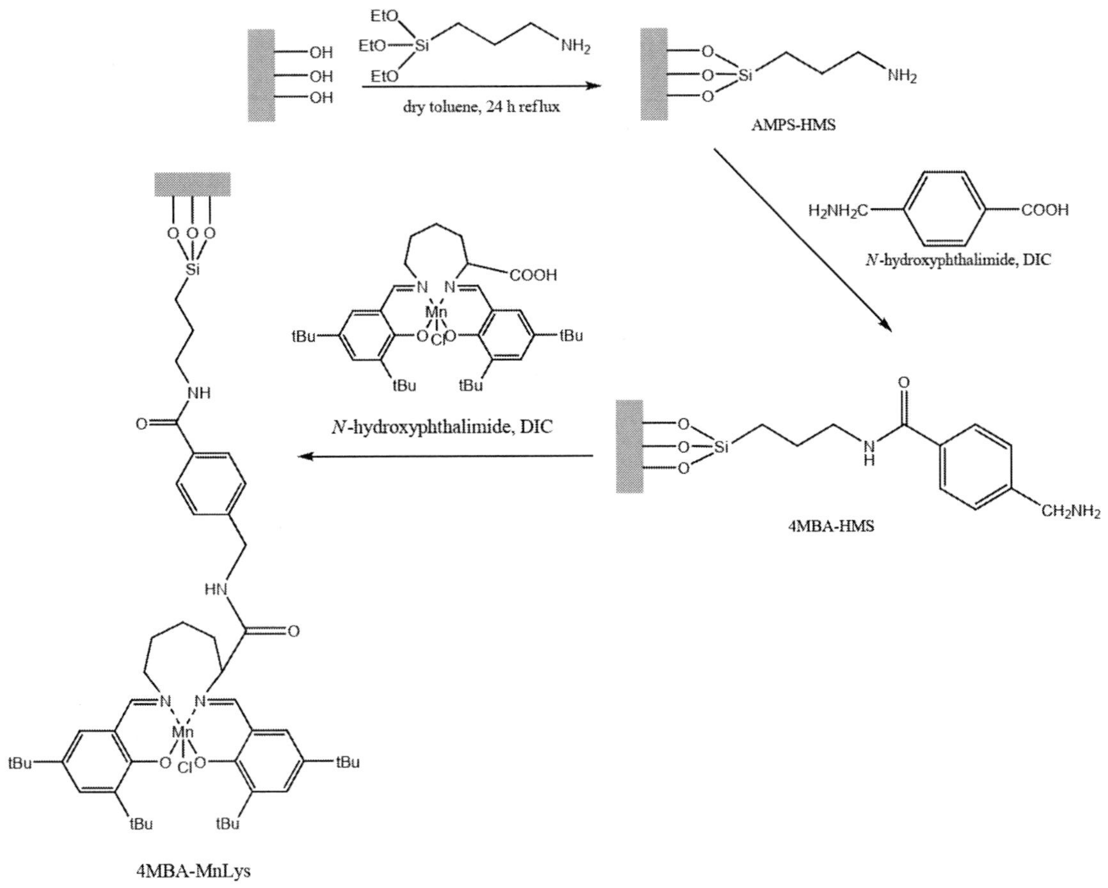

Scheme 1. The procedure for HMS modification and also the covalent attachment of Mn-Sal-Lys onto the functionalised silica.

2.6 Catalytic reactions

The procedure for the homogeneous reaction was adapted from literature [24]. Into a three necked round-bottomed flask equipped with a magnetic stirrer and condenser, 0.01 mmol of catalyst, 2.5 mL cyclohexene, 4.6 mL isobutyraldehyde and 50 mL acetonitrile were placed. The reaction was allowed to reach the desired temperature and air was then bubbled through the reaction. The reaction was monitored using gas chromatography until completion, typically 2 h. For the immobilised reaction, developed by Zhang and co workers [25], 0.03 g of catalyst, 20 mmol cyclohexene, 50 mmol isobutyraldehyde and 50 mL acetonitrile were loaded into an identical set-up as the homogeneous reaction. Air was bubbled through the reaction for 6 h. The reaction mixture was analysed with gas chromatography.

3. RESULTS AND DISCUSSION

3.1. Preparation characterisation of the catalysts

A standard procedure was used for the synthesis of Jacobsen's catalyst and the Mn-Sal-Lys was adapted form literature as described in the previous section. Fig. 1 represents the FT-IR for (A) Jacobsen's and (B) Mn-Sal-Lys complexes. The peaks at 2950 and 2863 cm^{-1} correspond to –CH2– stretch. The group of peaks between 1650 – 1400 cm^{-1} are assigned to $\nu_{C=N}$, ν_{C-N} and ν_{COO-} of the phenyl groups. Table 1 lists the results from CHN analysis.

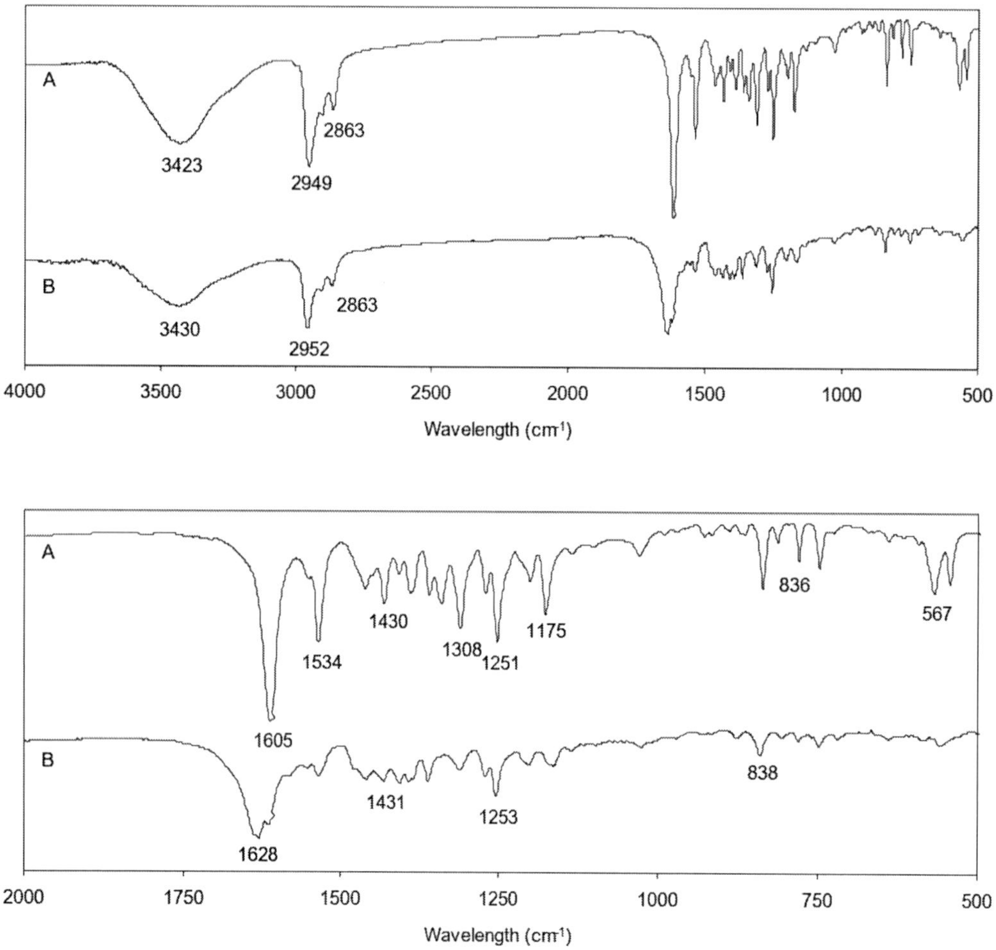

Figure 1. FT-IR spectra (A) Jacobsen's (B) Mn-Sal-Lys.

Table 1. CHN analysis of homogeneous complexes.

Sample	Expected			Found		
	%C	%H	%N	%C	%H	%N
Jacobsen	79.01	9.97	5.12	78.71	10.18	5.06
Mn-Sal-Lys	64.8	7.8	4.2	66.61	8.12	3.13

3.2. Characterisation of the silicate materials

Fig. 1 shows the adsorption/desorption BET isotherms for HMS, AMPS-HMS and AMPS-MnLys. The isotherms show the porosity of the material within the channels of the silicate support. Mesoporous media have pores in the range of 20-500 Å however it would be optimistic if there were no micropores in the silicates. The isotherms are a combination of Type I and Type IV isotherms [26]. The presence of a microporous phase may also be explained by monolayer coverage of the wall by the adsorbate [27]. The mesoporosity of the silicate is within the range of 0.1-0.5 Pi/Po, and defined by the height and steepness of the sorption slope. The higher and sharper the step the better defined the mesoporous structure. The hysteresis loop in the region > 0.4 Pi/Po suggests there is textural mesoporosity in all the silicates but they appear lower down the scale than expected. Textural porosity arises from the aggregation of small fundamental particles through intergrowth or other associative mechanisms [28] and is independent of pore size. This demonstrates there is significant mesoporous framework pores and textural mesoporosity. A summary of the adsorption characteristics can be found in Table 2. It is obvious that the surface area for the AMPS-HMS is lower than the immobilised complex 4MBA-MnLys. It would appear that solvent extraction is an inferior method for template removal than calcination. However Zhang and co workers found that extending solvent extraction from 14 h to 48 h effectively removed the template [25]. Table 3 shows the XPS spectra for HMS and AMPS-HMS and demonstrates that the surface modification with aminopropyl-groups has been successful.

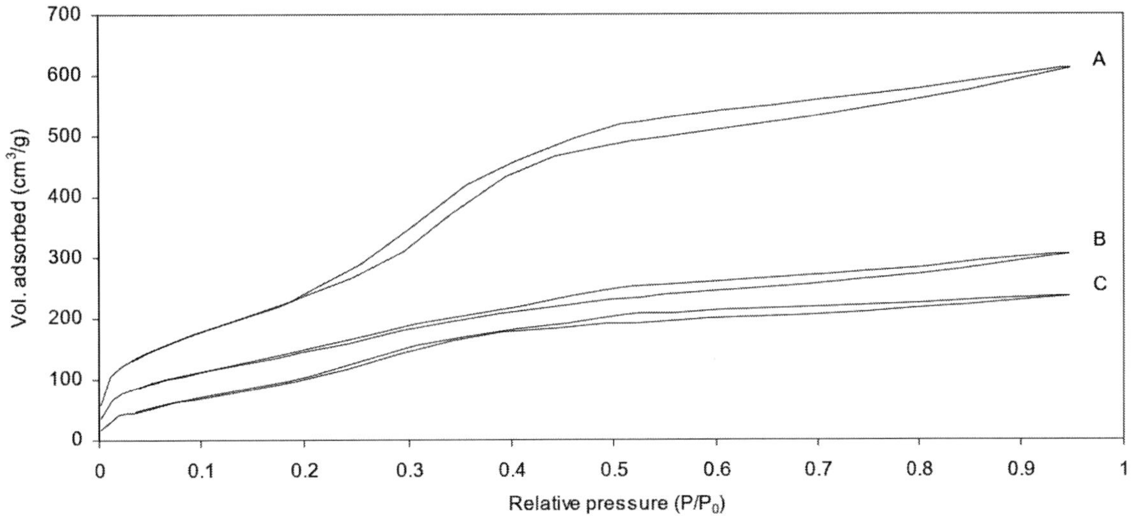

Figure 2. N_2 Adsorption-desorption isotherms for ethanol extracted S°I° HMS (A), AMPS-HMS (B) and (C) 4MBA-MnLys. P is the equilibrium pressure of the adsorbates and P_0 the saturated pressure of the adsorbates.

Table 2. Summary of the silicate adsorption properties.

Sample	Area $_{BET}$ m^2g^{-1}	Pore Volume mLg^{-1}	Pore size Å
HMS	771	1.162	24.8
AMPS-HMS	326	0.495	22.2
AMPS-MnLys	482	0.552	20.6

Table 3. XPS spectra for HMS and AMPS-HMS

Sample	**O 1s**	**C 1s**	**N 1s**	**Mn 2p**	**Cl 2p**	**Si 2p**
HMS	66.1%	9.0%	-	-	-	24.9%
AMPS-HMS	60.8%	12.0%	2.9%	-	-	24.3%

3.3. *Anchoring of Mn-Sal-Lys onto supports*

The homogeneous catalyst Mn-Sal-Lys was anchored onto the supports using covalent attachment of the carboxylic acid group of the lysine diimne bridge. The concentration of the complex on the silicate is low and hidden by the signals from the SiO_2 on the FT-IR. The peak at 1621 cm^{-1} corresponds to $\nu_{C=N}$ and $\nu_{C=O}$ envelope. The peaks at 1075 and 942 cm^{-1} represents the Si–O–Si and Si–OH.

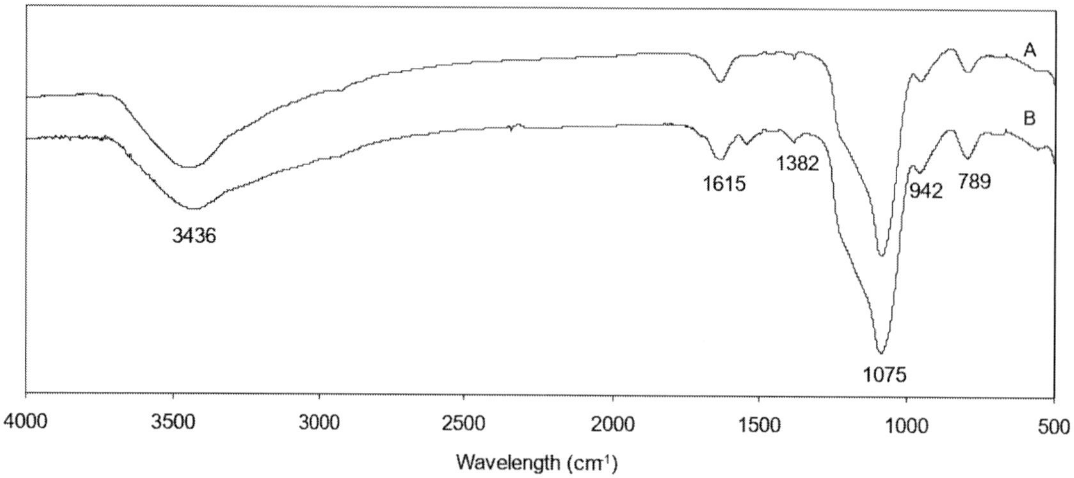

Figure 3 FT-IR of immobilised complexes of (A) AMPS-MnLys and (B) 4MBA-MnLys

3.3. *Catalytic experiments*

The catalytic experiments using the homogeneous and immobilised catalysts were for the aerobic epoxidation of cyclohexene and styrene with isobutyraldehyde as a sacrificial co reductant. One of the most impressive factors of this experimental procedure is the mild conditions under which it operates. The method of attachment at the diimine bridge was chosen as it has been reported that this can improve the %ee compare to attachment of the complex through the aldehyde fragments in the 5 position[19]. It is believed that the immobilisation at the diimine bridge changes the steric environment thus improving the %ee. Even though the %ee for styrene has not been extensively studied here it is hoped this will be explored in the future.

The Mn-Sal-Lys has similar activity to *Jacobsen* catalyst but shows a surprising increase in yield for the epoxidation of cyclohexene as shown in Table 4. Perhaps the less sterically hindered diimne bridge allows easier access of the substrate to the active site. Currently there is no comparison for the homogeneous complexes for the epoxidation of styrene. The results for the epoxidation of cyclohexene using *Jacobsen* catalyst are consistent with Zhang an co workers with the reaction at 35 °C providing the best results[25].

As expected the homogeneous phase is more active than the heterogeneous phase. Unusually the yield of cyclohexene oxide is slightly higher than the homogeneous complexes. This is counter-intuitive and there is no clear explanation for this behaviour. However the biggest advantage of using AMPS-MnLys and 4MBA-MnLys is their reuse. Between each run the catalyst was removed from the reaction mixture using vacuum filtration and washed several times with acetonitrile and diethyl ether. It was reloaded into the reactor and recharged with more substrate and co reductant. Both AMPS-MnLys and 4MBA-MnLys show good reproducibility of results even after 4-5 runs. An important factor in heterogeneous catalysis is to test for leaching. No results are presented here however it is being tested at the moment.

Table 4. Epoxidation at varying temperatures of alkenes catalysed by the homogeneous and immobilsed mnanganese(III) salen complexes

Catalyst	Alkene	T (°C)	Run	t (h)	Conv. (%)	Yield (%)
Jacobsen	Cyclohexene[a]	25		2	95	81
	Cyclohexene[a]	35		1	97	82
	Cyclohexene[a]	45		1	>99	73
	Styrene[b]	25		24	72	50
Mn-Sal-Lys	Cyclohexene[a]	25		2	>99	88
	Cyclohexene[a]	35		1	>99	86
AMPS-MnLys	Cyclohexene[c]	25	1	6	>99	93
	Cyclohexene[c]	25	2	6	>99	95
	Cyclohexene[c]	25	3	6	>99	96
	Cyclohexene[c]	25	4	6	>99	94
	Cyclohexene[c]	25	5	6	55	53
	Cyclohexene[c]	25	6	6	>99	93
	Cyclohexene[c]	25	7	6	75	71
4MBA-MnLys	Cyclohexene[c]	25	1	6	>99	90
	Cyclohexene[c]	25	2	6	>99	92
	Cyclohexene[c]	25	3	6	>99	93
	Cyclohexene[c]	25	4	6	93	84

[a] Experimental conditions: 0.01 mmol of catalyst, 2.5 mL cyclohexene, 4.6 mL isobutyraldehyde and 50 mL acetonitrile

[b] Experimental conditions: 0.01 mmol of catalyst, 2.8 mL styrene, 4.6 mL isobutyraldehyde and 50 mL acetonitrile

[c] Experimental conditions: 0.03 g of catalyst, 20 mmol cyclohexene, 50 mmol isobutyraldehyde and 50 mL acetonitrile

4. Conclusion

A new salen-type complex has been successfully synthesised and covalently anchored onto mesoporous silicate, HMS *via* a peptide-type bond between the diimine bridge of the complex and the silicate. Two different functional groups have been used to modify the surface of HMS with (1) aminopropyl and (2) 4-(aminomethyl)benzoic acid.

The immobilised complexes, AMPS-MnLys and 4MBA-MnLys, along with two homogeneous catalysts, *Jacobsen* and Mn-Sal-Lys were used in the aerobic oxidation of styrene and cyclohexene with isobutyraldehyde acting as a sacrificial reductant. As expected the homogeneous complexes were more active than the heterogeneous catalysts allowing for a good selectivity for cyclohexene oxide of *ca.* 80 % however the immobilised complexes show very good selectivity and reproducibility of results even after several runs. More work needs to be carried out to test for leaching of the immobilised complexes and to see if linker length has an effect on the enantioselectivity of styrene.

ACKNOWLEDGMENTS

Special thanks to Engineering and Physical Sciences Research Council (EPSRC) for funding this work through project reference EP/D038162/1.

REFERENCES

1. Swern, D., T.W. Findley, and J.T. Scanlan, *Epoxidation of oleic acid, methyl oleate and oleyl alcohol with perbenzoic acid.* 1944. **66**: p. 1925-1927.
2. Findley, T.W., D. Swern, and J.T. Scanlan, *Epoxidation Of Unsaturated Fatty Materials With Peracetic Acid In Glacial Acetic Acid Solution.* 1945. **67**(3): p. 412-414.
3. Wong, D.T. and L.C. Terriere, *Epoxidation Of Aldrin Isodrin And Heptachlor By Rat Liver Microsomes.* 1965. **14**(3): p. 375-&.
4. Ray, J.W., *Epoxidation Of Aldrin By Housefly Microsomes And Its Inhibition By Carbon Monoxide.* 1967. **16**(1): p. 99-&.
5. Schonbro.Rd, et al., *Epoxidation Of Aldrin By Microsomes From Fc Strain Of Housefly - Genetic Survey.* 1973. **3**(3): p. 259-270.
6. Friedman, P.A. and M.W. Smith, *Epoxidation Of Several Vitamins-K By Rat-Liver Microsomes.* 1979. **28**(6): p. 937-938.
7. Groves, J.T., T.E. Nemo, and R.S. Myers, *Hydroxylation and Epoxidation Catalyzed by Iron-Porphine Complexes - Oxygen-Transfer from Iodosylbenzene.* Journal of the American Chemical Society, 1979. **101**(4): p. 1032-1033.
8. Groves, J.T. and R.S. Myers, *Catalytic Asymmetric Epoxidations with Chiral Iron Porphyrins.* Journal of the American Chemical Society, 1983. **105**(18): p. 5791-5796.
9. Groves, J.T. and T.E. Nemo, *Epoxidation Reactions Catalyzed by Iron Porphyrins - Oxygen-Transfer from Iodosylbenzene.* Journal of the American Chemical Society, 1983. **105**(18): p. 5786-5791.
10. Irie, R., et al., *Catalytic Asymmetric Epoxidation of Unfunctionalized Olefins.* Tetrahedron Letters, 1990. **31**(50): p. 7345-7348.
11. Zhang, W. and E.N. Jacobsen, *Asymmetric Olefin Epoxidation with Sodium-Hypochlorite Catalyzed by Easily Prepared Chiral Mn(Iii) Salen Complexes.* Journal of Organic Chemistry, 1991. **56**(7): p. 2296-2298.
12. Zhang, W., et al., *Enantioselective Epoxidation of Unfunctionalized Olefins Catalyzed by (Salen)Manganese Complexes.* Journal of the American Chemical Society, 1990. **112**(7): p. 2801-2803.
13. Yoon, H. and C.J. Burrows, *Catalysis of Alkene Oxidation by Nickel Salen Complexes Using Naocl under Phase-Transfer Conditions.* Journal of the American Chemical Society, 1988. **110**(12): p. 4087-4089.
14. Samsel, E.G., K. Srinivasan, and J.K. Kochi, *Mechanism of the Chromium-Catalyzed Epoxidation of Olefins - Role of Oxochromium(V) Cations.* Journal of the American Chemical Society, 1985. **107**(25): p. 7606-7617.
15. IUPAC, *Manual of Symbols and Terminology.* Pure and Applied Chemistry, 1972. **31**: p. 578-638.
16. Tanev, P.T. and T.J. Pinnavaia, *Mesoporous silica molecular sieves prepared by ionic and neutral surfactant templating: A comparison of physical properties.* Chemistry of Materials, 1996. **8**(8): p. 2068-2079.
17. Kim, G.J. and S.H. Kim, *Immobilization of new Mn(salen) complex over MCM-41 and its activity in asymmetric epoxidation of styrene.* Catalysis Letters, 1999. **57**(3): p. 139-143.
18. Benedito, F.L., et al., *Study of metalloporphyrin covalently bound to silica as catalyst in the ortho-dianisidine oxidation.* Applied Catalysis a-General, 2003. **250**(1): p. 1-11.
19. Silva, A.R., et al., *Covalent attachment of chiral manganese(III) salen complexes onto functionalised hexagonal mesoporous silica and application to the asymmetric epoxidation of alkenes.* Microporous and Mesoporous Materials, 2006. **91**(1-3): p. 128-138.
20. Zhang, H. and C. Li, *Asymmetric epoxidation of 6-cyano-2,2-dimethylchromene on Mn(salen) catalyst immobilized in mesoporous materials.* 2006. **62**(28): p. 6640-6649.
21. Yamada, T., et al., *Highly Efficient Method for Epoxidation of Olefins with Molecular Oxygen and Aldehydes Catalyzed by Nickel(II) Complexes.* Chemistry Letters, 1991. **20**(1): p. 1.
22. Deng, L. and E.N. Jacobsen, *A Practical, Highly Enantioselective Synthesis of the Taxol Side Chain via Asymmetric Catalysis.* J. Org. Chem., 1992. **57**(15): p. 4320-4323.
23. Wang, R.-M., et al., *Amino acid Schiff base complex catalyst for effective oxidation of olefins with molecular oxygen.* Journal of Molecular Catalysis A: Chemical, 1999. **147**(1-2): p. 173.
24. Zhao, J., J. Han, and Y. Zhang, *Preparation of encapsulated and anchored alanine–salicylaldehyde Schiff base Mn(III) (Sal–Ala–Mn) complexes by sol–gel method and their performance in aerobic epoxidation of cyclohexene.* Journal of Molecular Catalysis A: Chemical, 2005. **231**(1-2): p. 129-135.
25. Zhang, Y.R., et al., *Manganese (III) salen complex anchored onto MCM-41 as catalyst for the aerobic epoxidation of olefins.* Microporous and Mesoporous Materials, 2006. **94**(1-3): p. 159-165.
26. Schneider, P., *Adsorption-Isotherms Of Microporous Mesoporous Solids Revisited.* 1995. **129**(2): p. 157-165.
27. Kresge, C.T., et al., *Ordered Mesoporous Molecular-Sieves Synthesized By A Liquid-Crystal Template Mechanism.* Nature, 1992. **359**(6397): p. 710-712.
28. Pauly, T.R., et al., *Textural mesoporosity and the catalytic activity of mesoporous molecular sieves with wormhole framework structures.* 1999. **121**(38): p. 8835-8842.

69e Utilisation of Heterogeneous Catalysts for the Selective Synthesis of Di- and Triglycerol

Arno Behr and **Falk Lindner**. Department of Biochemical and Chemical Engineering, Technical University of Dortmund, Emil-Figge-Str. 66, 44227 Dortmund, Germany

Glycerol becomes more and more interesting because of the surplus arising from the biodiesel production. Thus the chemical industry looks for advanced utilisation of glycerol. Telomers could be used as surfactants whereas oxidation and carbonylation could lead to interesting chemicals via new catalytic pathways. In this contribution the focus is on the oligomerisation of glycerol. Di- and triglycerol, which are the main products, are of increasing interest especially for body care products or as starting materials for polyglycerol esters, which can be applied as emulsifiers. The following scheme shows the reactions taking place. Scheme 1: Oligomerisation of glycerol By today glycerol oligomers are produced homogeneously catalysed using alkali hydroxides as catalysts. Such catalysts reveal two big disadvantages. The first one is the fact that the catalyst is homogeneously solved in the reaction mixture. To get rid of the catalyst expensive separation steps have to be used, for example distillation. The second problem arising through utilisation of homogeneous catalysts is the product spectrum of the reaction. Short oligomers such as di- and triglycerol are of large interest whereas the utilisation of higher oligomers is not so interesting. The selectivity of homogeneous catalysts is by far too bad. This problem can be handled using heterogeneous catalysts. Their separation is easy and their selectivity towards di- and triglycerol is quite good. However, so far heterogeneous catalysts are not as active as homogeneous catalysts but much more selective. The presentation deals with the description of appropriate heterogeneous catalysts. Information is given about the conversions as well as the selectivity at optimum reaction conditions. Moreover, experiments under continuous conditions were performed and their results will also be shown.

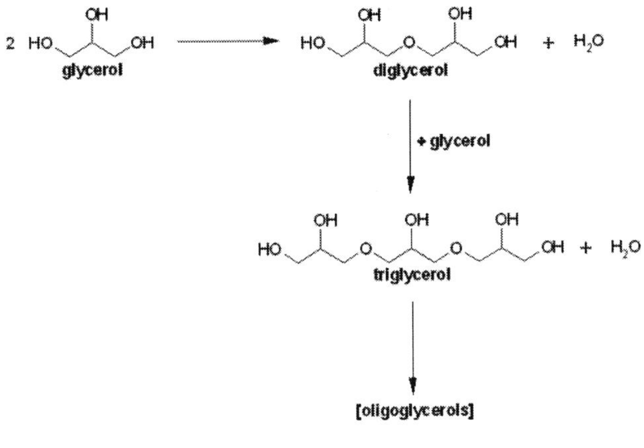

Hydrogen Conversion by Reformation of Jet Fuel in a Supercritical Water Medium – Effects of the Water-to-Fuel Feed Ratio on the Conversion

Satya Putta, Jason Picou, Jonathan Wenzel and Sunggyu Lee, Missouri University of Science and Technology, Rolla, MO

Abstract

The noncatalytic direct reformation of jet fuel was studied in a supercritical water medium using a specially designed 400-mL Haynes® 230 alloy tubular reactor. The experiments were conducted at 1042±5 K, 240.9±0.1 bar, and at water flow rates of 7.5, 15, 30 g/min with varying water-to-fuel feed ratio of 7.5, 15, 30. The principal reactions are reformation of jet fuel into hydrogen and carbon oxides and pyrolysis of jet fuel into lighter hydrocarbons and coke. The effluent gas contained H_2, CO, CO_2 and CH_4, with smaller concentrations of C_2H_6. This paper discusses the effects of the water-to-fuel ratio on the hydrogen and hydrocarbon percentage in the effluent gas. These percentages reflect the activity of the reformation and pyrolysis reactions. The results provide vital information for the optimization of this novel reformation process.

Introduction

Hydrogen produced from liquid hydrocarbon fuels may become important, since it can provide us with the capability of on-site generation of hydrogen without requiring massive infrastructure. In this paper, direct noncatalytic reformation of jet fuel in a supercritical water medium is studied. Jet fuel consists of branched and cyclic hydrocarbons, which require high energies to break the bonds, and contains a high sulfur content, which acts as a poison to most of the catalysts used in catalytic reformation processes. The noncatalytic reformation of jet fuel to hydrogen is conducted in a supercritical water medium, which functions as a highly energized reforming agent and also as a homogenizing reaction medium. The noncatalytic reformation and its process efficiency are not hindered by the sulfur content of the jet fuel due to the total absence of any noble metal catalyst. At supercritical conditions, water exhibits both liquid and gas-like properties, at which it is no longer a polar solvent and can easily dissolve a large variety of organic species resulting in homogeneous supercritical mixtures. In this paper, the effect of water-to-fuel feed ratio on the process conversion as well as its principal reactions is experimentally examined under well-defined experimental conditions of the supercritical water reformation.

Chemical Reactions

The two principal reactions in the reformation of jet fuel are the hydrocarbon reformation reaction and the pyrolysis reaction. The overall reformation reaction of jet fuel may be expressed as[2]:

$$C_{12}H_{26} + 12\ H_2O \rightarrow 12\ CO + 25\ H_2 \qquad \Delta H^{\circ}_{298} = 1866.2\ \text{kJ/mol} \qquad (1)$$

In reaction (1), jet fuel is represented and modeled as n-dodecane for stoichiometric simplicity as explained later. Jet fuel is also competitively pyrolyzed simultaneously by the following reaction:

$$C_{12}H_{26} \rightarrow C_aH_b + C_xH_y + wH_2 \qquad (12 = a + x \text{ and } 26 = b + y + 2w) \qquad (2)$$

The pyrolysis reaction produces unsaturated, hydrogen-deficient, hydrocarbons, which may undergo cyclization polymerization eventually, if left uncontrolled, producing coke.

Apparatus and Chemicals

Figure 1 is a schematic process flow diagram of the supercritical water reformation system. The system comprises liquid reactant feed system, integrated heat exchanger, reactor vessel, reactor heaters/insulators, product sample collection system, and PC-based data acquisition and control system. The reactor is constructed of Hanes® 230 alloy with an internal reactor volume of 383 mL, which permits supercritical water experiments to be conducted under up to 36.3 MPa at 800°C.[1] The product sampling system allows collection of effluent gas and liquid samples for compositional analysis and for closure of material balances. Analysis of effluent gas was conducted using a HP 5890 Series A gas chromatograph with a thermal conductivity detector. It is calibrated to detect H_2, CO, CH_4, CO_2, C_2H_2, C_2H_4, and C_2H_6.

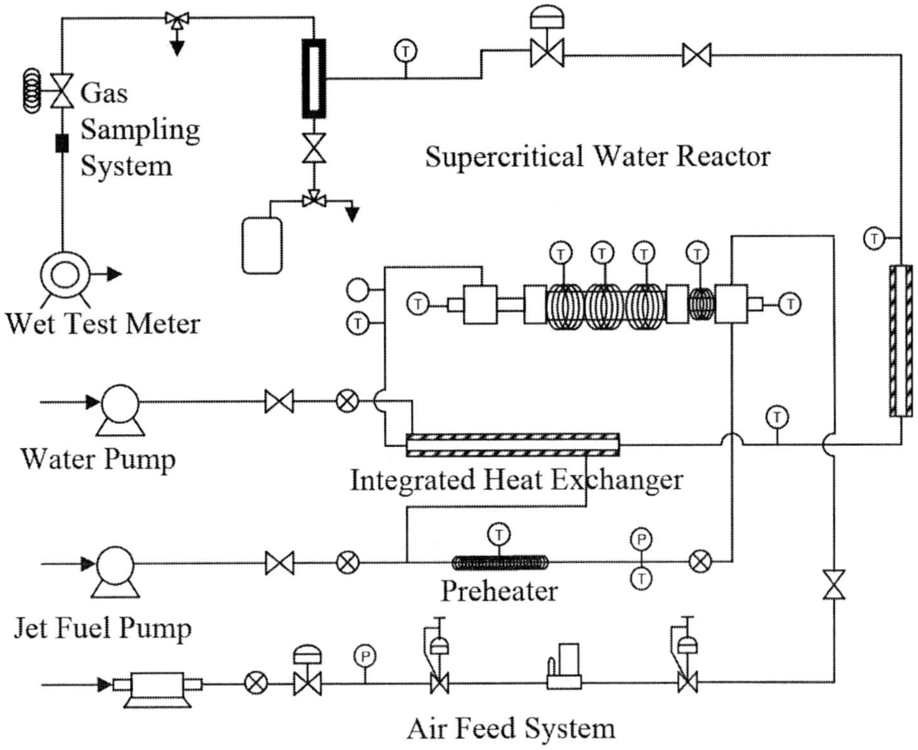

Figure 1. A schematic of the supercritical water reformation system

The water fed into the reactor was deionized water. The jet fuel used for this work was both civilian jet fuel (Jet-A) and military aviation logistic fuel (JP-8), both of which are a mixture of cyclic, chain, and branched hydrocarbons. An ASTM D2887 boiling range distribution analysis determined that the average carbon number for both Jet-A and JP-8 was 12. Therefore, for stoichiometric simplicity, both fuels are often represented and modeled as n-dodecane with molecular formula of $C_{12}H_{26}$.[2, 3]

Experimental

The water-to-fuel feed ratio was varied to investigate the effects upon the ambient reactor effluent gas composition. The experimental conditions were maintained at 1042 K and 240.9 bar. The water-to-fuel feed ratio was varied by controlling the amount of water and fuel fed into the reactor system through the liquid feed system. The water-to-fuel ratio is calculated as ratio of mass flow rate (grams per minute) of water fed into the reactor system to the grams of fuel fed per minute. Three different water-to-fuel ratios of 7.5, 15, 30 were investigated at 1042±5 K, 240.9±0.1 bar at water flow rates of 7.5, 15 and 30 g/min. The temperature and pressure conditions for the experiments were chosen from a region of technological significance for this specific reformation reaction process[2,3].

Results and Discussion

Table 1 shows the run conditions and the resultant effluent gas compositions. The hydrocarbon percentage is the mole percent of the effluent gas that essentially includes methane, ethane or ethene. The percentages reported are calculated on mole percentages on a dry basis.

Table 1. Experimental run matrix and effluent gas composition

Run ID	Water Flow (g/min)	Fuel Flow (g/min)	Water to Fuel Ratio	Mol-% hydrogen in effluent	Mol-% of total hydrocarbon gas in the effluent
1	7.5	1.00	7.5	35.60	44.00
2	7.5	0.50	15.0	44.03	30.60
3	7.5	0.25	30.0	49.80	21.77
4	15.0	2.00	7.5	27.02	55.47
5	15.0	1.00	15.0	34.27	47.84
6	15.0	0.50	30.0	45.79	27.62
7	30.0	4.00	7.5	28.56	56.17
8	30.0	2.00	15.0	30.13	51.05
9	30.0	1.00	30.0	44.56	33.13

First, the experiments were performed at constant water flow rate of 7.5 g/min with different flow rates of jet fuel under otherwise nominally same reactor conditions, i.e., 1042 K and 240.9 bar. Figure 2 illustrates, with an increase in water-to-fuel ratio the percentage of hydrogen in the effluent stream increased from 35.6 to 49.8 mol-%, which could be due to the reformation reaction becoming more active with higher water-to-fuel ratios. It is assumed that any hydrocarbon gas molecules contained in the effluent gas are due to pyrolysis. As the reformation reaction becomes more active on a competitive basis, fewer light hydrocarbons are produced by the pyrolysis reaction. The hydrocarbon mole percentage in the effluent gas stream decreased from 44 to 21.77 mol-% as shown in figure 2.

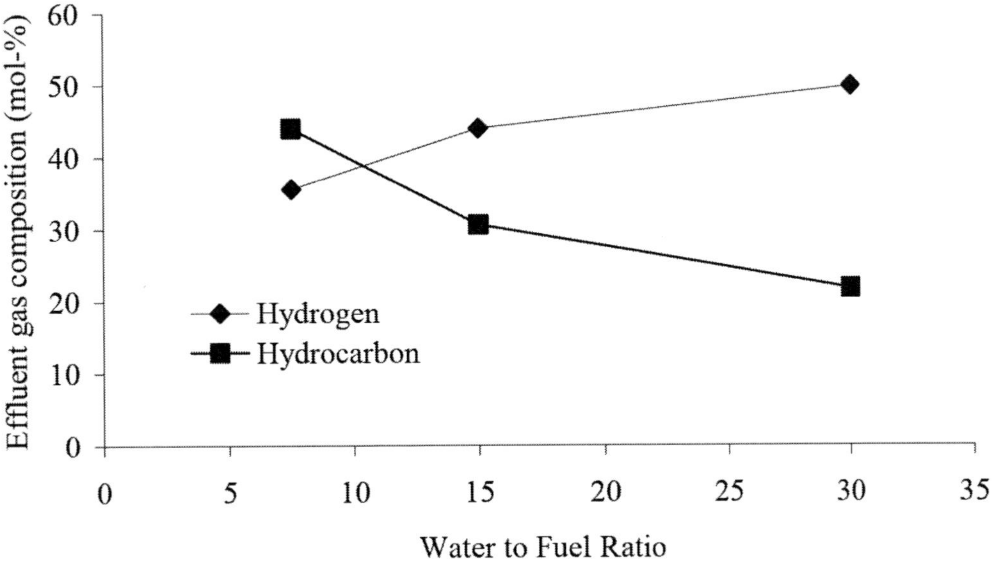

Figure 2. Hydrogen and hydrocarbon mole percents in the ambient effluent gas as a function of water-to-fuel ratio. Experimental conditions: T=1042±5 K, P=240.9±0.1 bar, Water feed rate = 7.5 g/min.

Similar experiments were conducted at a water flow rate of 15 g/min as illustrated in Figure 3. The hydrogen mole percentage in the effluent gas increased from 27.02 to 45.79 mol-%. As in the earlier case of 7.5 g/min, a significant increase in the hydrogen percentage was noticed. The hydrocarbon gas mole percentage decreased from 55.4 to 27.62 mol-%.

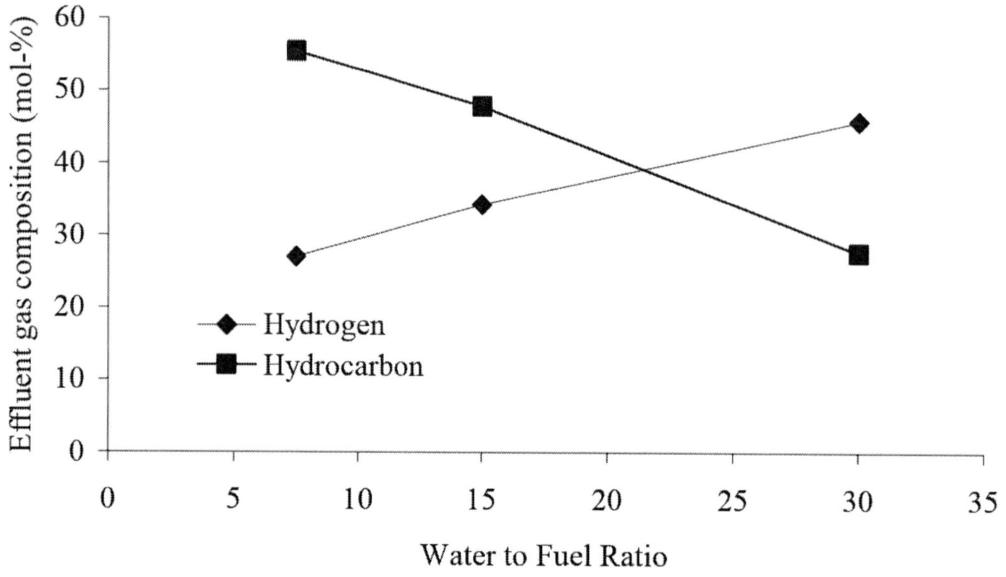

Figure 3. Hydrogen and hydrocarbon mole fractions in the ambient effluent gas as a function of water to fuel ratio. Experimental conditions: T=1042±5 K, P=240.9±0.1 bar, Water feed rate = 15 g/min.

The experiments were repeated with a higher water flow rate of 30 g/min, the results of which can be seen in Figure 4. The hydrogen mole percentage in the effluent gas increased from 28.56 to 44.56%, while hydrocarbon gas in the effluent stream decreased from 56.17 to 33.13%.

All three sets of experiments support the opinion that with increases in the water-to-fuel ratio, the reformation reaction becomes more active than the pyrolysis reaction thus increasing the resultant hydrogen percentage in the effluent gas. The effects of higher water-to-fuel ratios in the reactor systems involve the following thermodynamic and kinetic significances that in turn affect the comparative reactivities of the involved reactions, i.e., reformation and pyrolysis. First, the effects of the higher water-to-fuel ratios in the current experiments were consistent irrespective of the reactor residence times investigated. Higher water feed rate runs in the current study represent shorter reactor residence times, and vice versa. Secondly, the higher water-to-fuel ratios ensure that the reactor condition will be maintained under supercritical mixture conditions longer in the duration than its counterpart lower ratios. This is due to the changing supercriticality of the reactor fluid mixtures as the reaction proceeds along the reactor length due to generation of reformation product gas molecules as well as appearance of pyrolyzed lower hydrocarbon fractions. Thirdly, the direct reformation reaction of the jet fuel hydrocarbon molecules may be limited by chemical equilibrium at the conditions investigated in this study. In such a case, the kinetic expression for the reformation reaction must include terms that are taking the chemical reaction equilibrium into account.

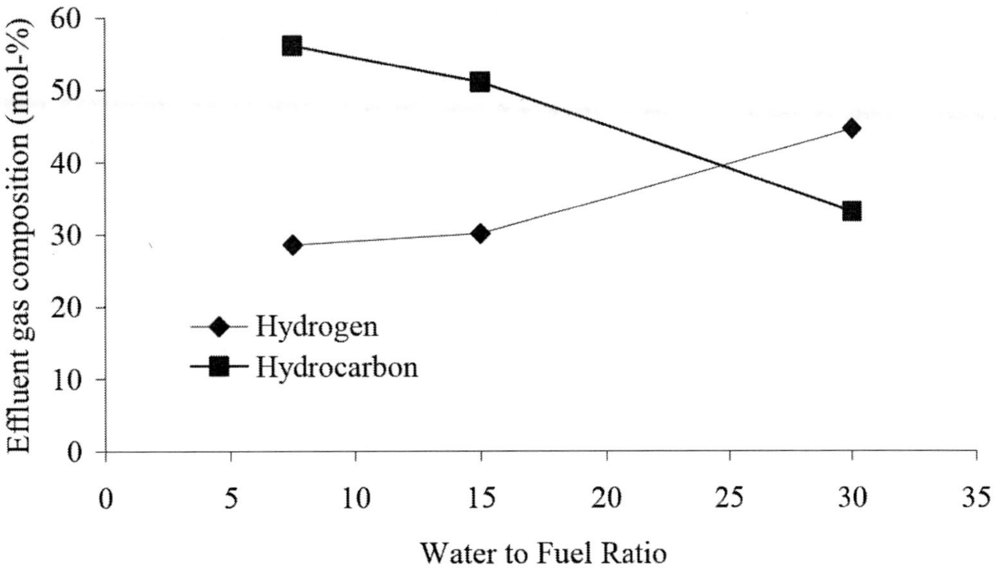

Figure 4. Hydrogen and hydrocarbon mole fractions in the ambient effluent gas as a function of water-to-fuel ratio. Experimental conditions: T=1042±5 K, P=240.9±0.1 bar, Water feed rate = 30 g/min.

Conclusion

The reformation of jet fuel was studied in a noncatalytic 0.4-L Haynes® Alloy 230 tubular flow reactor. Three different water-to-fuel ratios of 7.5, 15, 30 were experimentally compared at 1042±5 K, 240.9±0.1 bar at water flow rates of 7.5, 15 and 30 g/min, respectively. The hydrogen percentage in the effluent product gas stream increased with increasing in water-to-fuel ratios for all water flow rates investigated in the current study. This is indicative of the reformation reaction becoming more active than the pyrolysis reaction with increasing water-to-fuel ratio. The thermodynamic and kinetic viewpoints on the experimental results are also presented.

Acknowledgements

The current project was sponsored by the U.S. Army under PE Number 0602705A through a subcontract from DRS Technical Services, Inc. The authors are grateful for insightful technological guidance and supervision provided by Dr. Terry Dubois, CERDEC, U.S. Army, Ft. Belvoir, Virginia.

References

1. Wenzel, J., Picou, J., Factor, M., Lee, S., "Kinetics of Supercritical Water Reformation of Ethanol to Hydrogen", presented at Materials Science & Technology 2007 Conference and Exhibition, Detroit, MI, September 2007.
2. Lee, S., Lanterman, H., Wenzel, J., Edwards, N., Adams, A., Wootton, J., Garcia, A., "Noncatalytic direct reformation of JP-8 fuel in supercritical water". Preprints-American Chemical Society, Division of Petroleum Chemistry, 2006, 51, 521-523.

3. Picou, J., Wenzel, J., Bryan Lanterman, H., Lee, S., "Autothermal Non-catalytic Reformation of Jet Fuel in a Supercritical Water Medium", presented at the AIChE 2007 Annual Meeting, Salt Lake City, Utah, November 4-9, 2007.

NEEDLE IN A HAYSTACK CATALYSIS - AN EXPERIMENTAL STUDY USING TEMPORAL ANALYSIS OF PRODUCTS (TAP) AIChE 2008 SPRING NATIONAL MEETING

Xiaolin Zheng, Washington University in St. Louis, St. Louis, MO, USA
John T. Gleaves, Washington University in St. Louis, St. Louis, MO, USA
Gregory S. Yablonsky, Saint Louis University, St. Louis, MO, USA
Thomas Brownscombe, Shell Oil Company, Houston, TX, USA
Anne Gaffney, ABB Lummus Global, Inc., Bloomfield, NJ, USA
Mike Clark, Rohm and Haas, Spring House, PA, USA
Scott Han, Rohm and Haas, Spring House, PA, USA

Introduction

Two general experimental approaches using different forms of catalytic materials are used in heterogeneous catalysis research. The "model catalyst" approach employing materials with well-defined surfaces, such as metal single crystals, provides a direct method of linking surface structure to elementary steps in a catalytic process.

The second approach, which can be called the "practical catalyst" approach, is the basis for most industrial catalyst development. Practical catalysts include supported metals, mixed metal oxides, and zeolites. The "practical catalyst" approach involves the systematic synthesis, testing, and characterization of numerous catalyst samples. In general, practical catalysts have complex ill-defined surfaces, and the active surface structure cannot be directly identified. Also, the measurement of catalytic activity under process conditions does not provide a direct link between structure and activity. In addition, the composition of a practical catalyst can change under the influence of the reaction medium resulting in a change in its properties. Consequently, practical catalyst development is hampered by a lack of fundamental information on the composition and structure of the catalytic phase or catalytic site.

Information from model catalyst studies is intended to provide guidance in the development of practical catalytic materials. However, the application of fundamental information has not been straightforward, since there is a significant "gap" between the conditions of "real world" catalytic processes, and those of model reaction studies. Bonzel [1] was the first to discuss the problem and coined the term "pressure gap" to emphasize the 10^{12} difference in reactant pressures. Practical and model catalyst experiments are also separated by a "materials gap" arising from the differences in structural and compositional complexity of catalyst samples. The materials gap presents a significant problem since the interplay of bulk and surface and the complex interactions of different constituents in practical catalysts are not easily reducible to a set of simple individual interactions. It is questionable whether results obtained at vacuum conditions on a set of well-defined uniform surfaces can be extrapolated to practical materials operating at atmospheric pressures, unless the uniform and practical samples can be directly compared experimentally.

Reactions on practical catalysts can be performed at vacuum conditions and at atmospheric pressures using a TAP (Temporal Analysis of Products) reactor system [2, 3, 4, 5]. This paper presents new results from a series of heterogeneous CO oxidation experiments using a single micron-sized catalyst particle surrounded by non-active quartz particles ("platinum needle in quartz haystack"). The catalyst particle occupies less than 1% of the cross-sectional area of the microreactor, so that the reaction zone can be considered a point source. This construction eliminates non-uniformity in reactant concentration and catalyst composition even in the domain of high conversions. The results are significant because they allow direct comparison between transient response experiments performed at vacuum conditions and steady flow experiments performed at atmospheric pressures – bridging the "pressure gap." Also, use of a single particle opens the door to a variety of new non-steady-state experiments that can supply intrinsic kinetic information, which can be directly related to catalyst composition and structure. For example, the single particle could be a well-defined single crystal, or an inert particle with metal deposited on the surface. Since all experiments use the same basic unit, a single particle, and the same reactor configuration and conditions, the results can be directly compared. Because TAP transient response experiments provide intrinsic kinetic data (e.g. number of active sites, site activation energy, etc.) and transport data, experiments performed on single crystals can be correlated with experiments performed on multi-crystalline materials. These experiments provide a new approach to bridging the materials gap.

Experimental

TAP Reactor

All single particle kinetic experiments were performed using a Temporal Analysis of Products (TAP-2) reactor system (Fig. 1A). The TAP-2 reactor can be used to perform TAP vacuum pulse response experiments, and atmospheric pressure steady-state, step-transient, and temperature programmed experiments [2, 3, 4, 5].

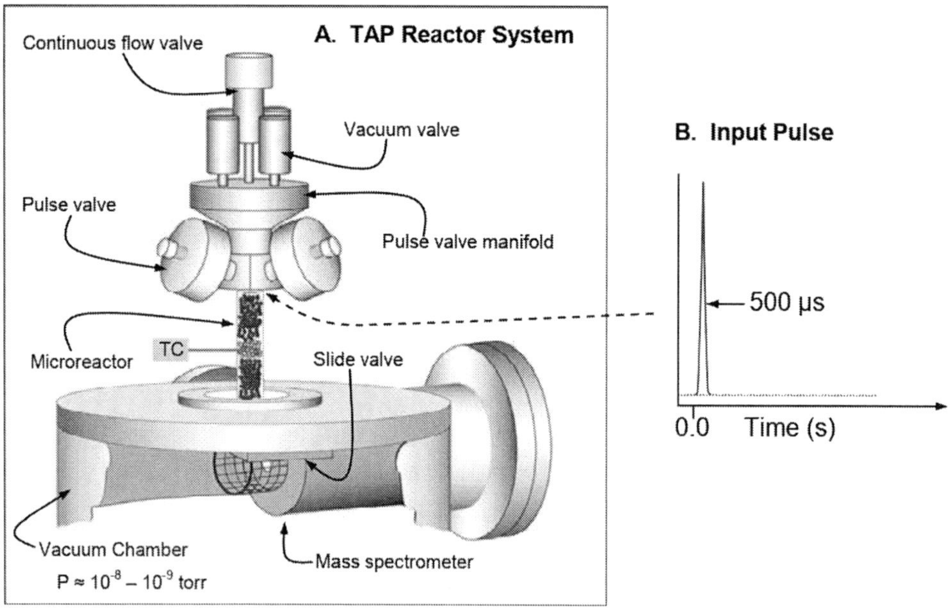

Fig. 1. (A) Schematic of TAP-2 reactor apparatus, which consists of 3 main parts: the pulse valve manifold, microreactor, and vacuum chamber containing the mass spectrometer. **(B)** Each inlet reactant pulse upon entering the microreactor is a very narrow pulse, often represented by the delta function. The pulse broadens as it travels through the microreactor, and in the case of an inert gas, the pulse shape will be the same as the standard diffusion curve.

Unlike conventional transient response experiments, TAP experiments use no carrier gas, and each pulse contains a very small amount of reactant ($\approx 10^{-9}$ mol/pulse). During a pulse, the gas density in the microreactor drops rapidly and the molecular mean free path becomes greater than the distance between particles. As a result, reactant and product molecules formed during chemical reaction move among the particles randomly, and independent of one another. Gas transport through the reactor can be characterized as Knudsen diffusion, which has the property that the diffusivity of each gas in a mixture is independent of the composition of the mixture as a whole [2, 3, 4, 5]. Mathematically, a Knudsen pulse can be described using Fick's second law (diffusion equation).

In a TAP Knudsen pulse response experiment, a narrow pulse of reactant gas (Fig. 1B) is injected into a packed-bed microreactor using a high-speed pulse valve. The microreactor is a stainless steel tubular reactor with dimensions 4.19 cm long and 0.64 cm in diameter. The reactor contains an internal thermocouple and can be operated isothermally or temperature programmed. The microreactor is attached to a high-throughput vacuum chamber (10^{-8}-10^{-9} torr) containing a UTI 100C QMS mass spectrometer. A 2-position slide valve assembly separates the reactor outlet and the mass spectrometer. When the slide valve is in the closed position, the microreactor can be operated at atmospheric pressures (atmospheric flow experiments). In the open position, the reactor operates at vacuum conditions (TAP pulse response experiments) [2, 3, 4, 5].

Catalyst Configuration

In a typical experiment, the microreactor was packed with approximately 100,000 inert quartz particles (210-250 μm in diameter) and a single catalyst particle (300-400 μm in diameter) usually positioned in the center of the reactor bed (Fig. 2A). The microreactor is heated resistively, and the internal temperature adjacent to the catalyst particle is sampled with a shielded thermocouple. Results presented in this paper are for particles of platinum (Pt) metal (Fig. 2B, 2C, 2D). The catalyst particle and the inert quartz particles are nonporous so that gas diffusion or mass transfer within particles does not occur.

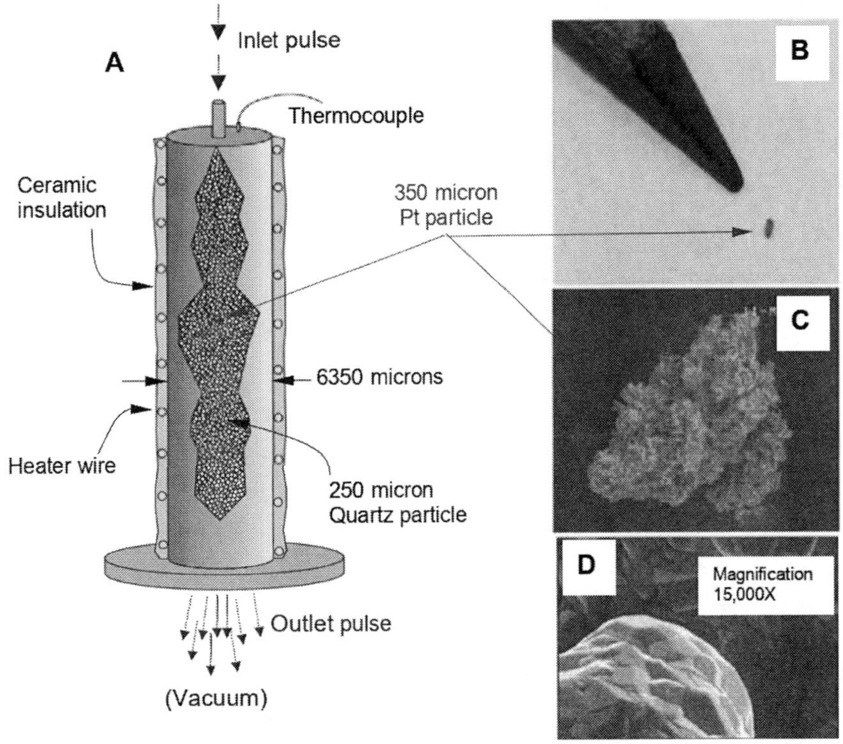

Fig. 2. **(A)** Schematic of TAP microreactor single particle configuration. The 350 μm diameter Pt particle is packed within a sea of inert quartz particles with diameters between 210-250 μm. There are approximately 100,000 inert particles and only 1 active particle in the entire microreactor volume. **(B)** Actual image of single particle Pt polycrystalline catalyst taken with a digital camera. The image compares the Pt particle to a pencil point to demonstrate the actual size of the catalyst particle. **(C)** SEM images showing the entire polycrystalline Pt particle. The diameter of the particle is estimated to be 350 μm. **(D)** Inset gives high magnification (15,000X) SEM image of Pt particle surface, which shows the surface is nonporous.

Results and Discussion

CO Oxidation on Pt Particle under TAP Vacuum Conditions

CO oxidation pulse response experiments were performed using the TAP "pump-probe" format. Mixtures of O_2/Ar (70/30) and CO/Ar (70/30) stored in separate pulse valves were injected into the microreactor forming a train of alternating pulses of O_2 and CO. In TAP experiments, argon is typically mixed with the reactant gases to act as a reference. Because it is inert, its transient response depends only on the transport process. It is used as a measuring stick to calculate diffusivities, conversion, and kinetic parameters. The pump-probe format maintains a relatively steady oxygen concentration on the catalyst by re-supplying any consumed oxygen in alternating O_2 pulses. In single Pt particle experiments, CO conversion was constant during multi-pulse pump-probe experiments at a steady reactor temperature. Therefore, based on experimental evidence, it can be stated that the oxygen coverage on the Pt particle is restored to the same value prior to each CO pulse.

In a pump-probe experiment [2, 3], the separation time between reactant pulses can be varied. Time delay between the reactant pulses can be changed to investigate the effects of changes in the surface lifetime of reactive adspecies on the dynamics of catalyst surface processes. In the experiments plotted in Figure 3, the O_2 and CO pulses were evenly spaced and separated by a time delay of 2 seconds. Identical experiments using time delays of 1 and 3 seconds were also performed, and no difference in the amount of CO_2 produced was observed. This indicates that the surface oxygen storage is approximately constant within the pump-probe time delays.

The amount of CO_2 produced in the pump-probe transient response experiment (Fig. 3A, 3C) can be determined by measuring the area (zeroth moment) under the transient response curve [6]. To determine CO_2 yield, the zeroth moment is normalized relative to the amount of CO in a single pulse. The absolute amounts of O_2, CO, and CO_2 are determined by comparing the response obtained in a reaction with one obtained using a standard blend. All responses are compared to the Ar response, which is the internal reference.

The black plot (Fig. 3C) represents the total CO_2 yield during one pump-probe cycle. The red and blue plots (Fig. 3C) show the CO_2 yield for O_2 and CO pulses respectively. The striking feature of these dependences is the high CO_2 yield that is achieved despite the catalyst particle occupies less than 0.1% of the total reactor volume. At 170 °C, the yield for the individual O_2 and CO pulses reach a maximum, making the total yield equal to approximately 95% during one pump-probe cycle. This means that under Knudsen diffusion conditions, at least 95% of the CO molecules pulsed into the reactor must strike the particle.

Above 170 °C, the CO2 yield decreases, more rapidly for the O_2 pulse than the CO pulse. From 170 to 350 °C, the CO_2 yield on the O_2 pulse decreases from a maximum of 52% to < 5%. In the same interval, the yield on the CO pulse decreases from 43% to 25%, so that the overall yield drops from 95% to less than 30%.

The difference between CO_2 dependences obtained in O_2 and CO pulse experiments can be explained within the modified adsorption mechanism (Langmuir-Hinshelwood mechanism). During the O_2 pulse, adsorbed oxygen may react with adsorbed CO molecules which are likely isolated at the reaction temperatures. In contrast, during the CO pulse, the adsorbed CO molecule may interact with different forms of stored oxygen, e.g. via the gradual destruction of the surface oxygen structure (surface phase islands). In this case, the CO_2 yield decreases not as rapidly. However, there is a general feature to both dependences. The yield peak ("turning point") occurs at the same temperature of 170 °C for both CO and O_2 pulses. Based on the adsorption mechanism, it is reasonable to assume that it is the point of transition from a CO dominated catalyst surface (below 170 °C) to an O_2 dominated catalyst surface (above 170 °C).

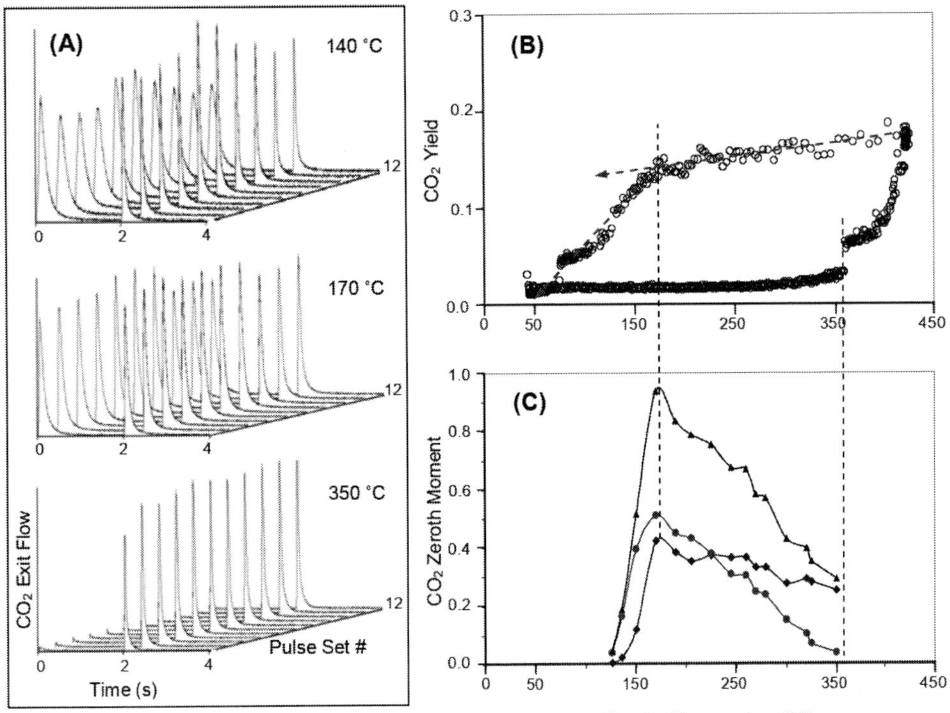

Fig. 3. Comparison of CO_2 produced during TAP vacuum pump-probe experiments and atmospheric flow experiments for CO oxidation over single Pt particle with the same composition of reactants. **(A)** A typical set of pump-probe CO_2 responses (m/e = 44) for reaction at 140, 170, and 350 °C. **(B)** CO_2 production observed from atmospheric flow experiment. The CO_2 produced while increasing reactor temperature is less than the CO_2 produced during reactor temperature decrease as shown by the counter-clockwise hysteresis loop. **(C)** CO_2 production observed from vacuum pump-probe experiment. The black line (-▲-) represents the total CO_2 yield. The red (-●-) and blue lines (-♦-) represent the CO_2 yield on the O_2 pulse and CO pulse, respectively.

CO Oxidation Experiments at Atmospheric Pressures

After completing the set of vacuum pulse response experiments, the slide valve was closed so that the reactor could be operated at atmospheric pressures. Prior to introducing an O_2/CO mixture, the particle bed was exposed to a hydrogen flow (20 cc/min diluted in Ar, H_2/Ar = 1) at 350 °C for 1 hour. The H_2 flow was used to remove any memory of the previous pump-probe experiments. The reactor was then cooled and a total flow of 50 cc/min of O_2, CO, and Ar (O_2/CO/Ar = 1) was introduced through the continuous valve, giving a gas residence time in the reactor of 1.8 s. Conversion under atmospheric conditions was quite high despite the short contact residence time inside the reactor (Fig. 3B).

The temperature dependence of CO_2 production was obtained by heating or cooling the reactor at a constant rate while maintaining an input flow of 50 cc/min. The internal reactor temperature was ramped from 40 °C to 430 °C over a 40 minute interval. Upon reaching 430 °C, the reactor temperature was held constant for 5 minutes and then decreased at the same ramp rate to room temperature.

During atmospheric pressure experiments, a small amount of the reactor effluent was diverted into the mass spectrometer chamber, and its mass spectrum was continuously monitored. The temperature dependence of the CO_2 yield is characterized by two branches, which form a counter-clockwise hysteresis loop (Fig. 3B). The lower branch occurs during the up ramp and the upper branch during the down ramp. On the lower branch of the hysteresis loop, at approximately 350 °C, a dramatic rise in CO_2 yield is observed. On the upper branch of the hysteresis loop, a maximum in CO_2 production of 20% is observed at 430 °C. At the temperature of approximately 170 °C ("turning point"), a significant decrease in CO_2 occurs.

The phenomenon of yield hysteresis is well known in heterogeneous catalysis, particularly in the heterogeneous CO oxidation reaction over noble metals [7]. Such behavior is also explained based on the adsorption mechanism in which O_2 and CO compete for active catalytic sites on the metal surface. From this point of view, the "turning point" is the temperature at which O_2 domination on the Pt surface is changed by the CO domination of the catalyst.

Comparison of Results from Vacuum and Atmospheric Pressure Regimes

Both vacuum and atmospheric pressure data exhibit several striking features. Perhaps most surprising is the very high conversion observed in TAP pulse response experiments. Despite the fact that the Pt particle occupies less than 0.1% of the total reactor volume, it is struck by at least 95 out of 100 CO molecules pulsed into the reactor.

In a TAP pulse response experiment, gas molecules move independently, primarily colliding with particles and the reactor walls. On average, a CO molecule experiences between 100,000 to 500,000 collisions before it exits the microreactor [8]. As a result, a molecule that initially bypasses the Pt particle can reverse its direction during its next collision and return to the Pt particle.

Conversion in atmospheric pressure experiments is also quite high. The residence time in the microreactor is longer than in a TAP experiment, but the apparent contact residence time in the reaction zone is about 2 orders of magnitude less. Nevertheless, CO conversion reaches 20%. Clearly, the real residence time in the reaction zone is larger, and mixing is also significant in atmospheric flow experiments.

An essential requirement for obtaining precise kinetic data is uniformity within the reaction zone in the reactant concentration profile and catalyst composition. The reactant concentration can be defined as uniform if its spatial difference is small compared to its maximum value. In single particle experiments the reaction zone is the particle surface. In vacuum pulse response experiments the reaction zone samples the entire reactor volume, and our estimates show that non-uniformity is negligible [9]. In atmospheric pressure experiments the reaction zone samples an area at least 67 times larger than the particle. Consequently, in both experiments non-uniformity is negligible, so it can be assumed the reaction zone is perfectly mixed.

Pulse response data obtained in high vacuum experiments can be used for direct determination of the number of active sites or storage of an active component [3, 10]. For a single particle, this number is comparable with the number of moles in a pulse, and can be determined in a single experiment. Data from TAP pulse response experiments can also provide details related to the steps in a complex reaction, i.e. kinetic coefficients of different reactions (reactant adsorption, interaction of reactants with other pre-adsorbed reactants), energies of activation and pre-exponential factors, and the dependence of all these parameters on the catalyst pre-treatment [3, 4, 10].

Conclusions

The new TAP reactor configuration, the single particle configuration, has demonstrated the high sensitivity of the TAP reactor and the ability to compare kinetic characteristics of the same catalyst under both high vacuum and atmospheric pressures. Experiments performed under vacuum and atmospheric pressures exhibit a "turning point" at 170 °C, which indicates a transition from the O2 dominating regime to the CO dominating regime (or vice versa) on the Pt single particle. In both cases, the "turning point" temperature is about the same at 170 °C. Therefore, the "turning point" temperature is the pressure independent characteristic which is governed by the steps of the complex catalytic processes with no participation of gas reactants or products (e.g. CO desorption, interaction between adsorbed oxygen and CO). The fact of observation of such pressure independent characteristic creates a new possibility for bridging across the pressure gap.

References

1. Bonzel, H.P. (1977) *Surface Science,* 68, pp. 236.
2. Gleaves, J.T., Ebner, J.R., Kuechler, T.C. (1988) *Catal. Rev.-Sci. Eng.,* 30, pp. 49-116.
3. Gleaves, J.T., Yablonskii, G.S., Phanawadee, P., Schuurman, Y. (1997) *Applied Catalysis A,* 160(1), pp. 55-88.
4. "The TAP Reactor in Catalysis: Recent Advances in Theory and Practice," (2007) *Catalysis Today,* ed. J. Pérez-Ramírez, E.V. Kondratenko, 121, pp. 1-124.
5. Yablonsky, G.S., Olea, M., Marin, G. (2003) *Journal of Catalysis,* 216(1-2), pp. 120-134.
6. Yablonskii, G.S., Shekhtman, S.O., Chen, S., Gleaves, J.T. (1998) *Ind. Eng. Chem. Res.,* 37(6), pp. 2193-2202.
7. Yablonskii, G.S., Bykov, V.I., Gorban, A.N., Elokhin, V.I. (1991), "Kinetic Models of Catalytic Reactions in Comprehensive Chemical Kinetics," ed. Elsevier, pp. 392.
8. Feres, R., Yablonsky, G.S. (2006) *Chemical Engineering Science,* 61, pp. 7864-7883.
9. Shekhtman, S.O., Yablonsky, G.S. (2005) *Ind. Eng. Chem. Res.,* 44(16), pp. 6518-6522.
10. Fushimi, R., Gleaves, J.T., Yablonsky, G.S., Gaffney, A., Clark, M., Han, S. (2007) *Catalysis Today,* 121(3-4), pp. 170-186.

205f Selectivity of Zirconia Supported Catalysts

Johnelle M. Halteman, Chemistry and Physics, Saint Mary's College, Notre Dame, IN 46556

The objective of this study was to synthesize zirconia supported catalysts and determine their selectivity in the production of hydrogen. Six different methods were used in synthesizing zirconia support using various zirconium precursors. The zirconia support was then loaded using impregnation and deposition-precipitation with a range of palladium, copper, and zinc precursor ratios as well as a range of palladium, copper, and zinc precusors. The zirconia supported catalysts were characterized using BET, XRD, Chemisorption, to measure the copper surface area, and a flow reactor, to determine the catalyst's selectivity. After characterization, it was found that the zirconia supported palladium, copper, and zinc catalysts were approximately as selective as non-zirconia supported palladium, copper, and zinc catalysts.

EXCEPTIONALLY ACTIVE SINGLE-SITE MESOPOROUS Pd/Al_2O_3 CATALYSTS FOR ALLYLIC ALCOHOL SELECTIVE AEROBIC OXIDATION

Dr. Adam F. Lee[1], Prof. Rik M. Brydson[2], Simon F.J. Hackett[1], Jamil Khan[3], Prof. Peter Styring[3] and Dr. Karen Wilson[1], (1) University of York, York, United Kingdom, (2)University of Leeds, Leeds, United Kingdom, (3) University of Sheffield, Sheffield, United Kingdom*

Abstract

The active site in the aerobic selective oxidation of allylic alcohols to their corresponding aldehydes over Pd/Al_2O_3 catalysts has been identified by combined operando X-ray studies and high resolution electron microscopy. The importance of isolated, electron-deficient Pd sites is shown.

Introduction

The heterogeneously catalyzed aerobic selective oxidation (selox) of hydrocarbons offers alternative, environmentally benign routes to a diverse range of valuable oxygenates for the fine chemical, pharmaceutical and agrochemical sectors.[1] These powerful catalytic technologies have the potential to replace stoichiometric inorganic oxidants such as Cr^{VI} salts or expensive homogeneous complexes (e.g. Mn salens), and their associated process disadvantages and safety issues.[2] Supported Platinum Group Metal particles are particularly promising partial oxidation catalysts[1,3], able to activate a range of alcohols and carbohydrates under mild conditions employing either neat substrate or green solvents including ionic liquids,[4] supercritical fluids[5] and even water.[6] The recent discoveries of Ru[7] and promoted Au catalysts[8] for such reactions is especially noteworthy, as these systems offer high turnover frequencies even during scale-up, although some require radical initiators and high air/O_2 partial pressures.[9] These reactions are highly regioselective (often >95 %) towards the aldehyde product, even in the presence of diverse functionalities including aryl, allylic groups. Most such studies in this area have utilised commercial catalyst formulations based upon high area amorphous carbon based or oxide supports. However the poor intrinsic performance of these materials relative to their homogeneous counterparts has often necessitated *ad hoc* promotion by non-noble metals to achieve even moderate yields.[10] Two key factors hamper the wider uptake and application of heterogeneous Pd selox catalysts amongst academic and industrial communities: first, uncertainty over the active site responsible for the rate-limiting oxidative dehydrogenation step;[11] and second, the use of conventional supports with restrictive pore dimensions that inhibit efficient mass/heat transfer to and from reaction sites and limit the range of viable solvents and substrates[12] In particular, recent studies have provided strong evidence that surface oxide plays a key role in enhancing the reactivity of supported palladium catalysts towards selox of allylic alcohols.[13] However, unequivocal identification of the active site in such Pd-derived systems has proven problematic, reflecting differential reaction conditions and rates of oxygen supply to the catalyst surface between different studies, and the difficulty in probing the active palladium species during such gas-liquid phase chemistry.

Since the energetics of metal clusters increasingly favour oxide terminated surfaces with decreasing cluster size, we hypothesized that a mesoporous high area support would serve to both stabilize highly dispersed palladium oxide nanoparticles, and facilitate efficient alcohol/aldehyde diffusion. Here we report the successful synthesis of tailored Pd-Al_2O_3 catalysts, and demonstrate that extremely low palladium loadings generate atomically dispersed Pd^{2+} surface species that confer exceptional selox activity of allylic alcohols.

Experimental

Catalyst Synthesis

Mesoporous alumina was prepared b the method of Vaudry and co-workers[14] from an aluminum hydroxide suspension obtained by hydrolysis of 87.5 g of aluminum sec-butoxide (Sigma-Aldrich >95 %) with 20.6 g of de-ionized water in 550 g of propan-1-ol. After 1 h stirring 21.6 g of lauric acid (Sigma-Aldrich 98 %) was added. The mixture was then aged (static) at room temperature for 24 h, and then heated for 48 h at 110 °C. The solid was then filtered and washed with ethanol and dried at room temperature. Template was removed by calcination under static air using a heating ramp of 25 ° C.min^{-1} to 250 °C, and then ramping to 550 °C at 5 °C.min^{-1} (note this reaction is highly exothermic and should be performed slowly). The sample was finally calcined at 550 °C for 2.5 h.

Supported palladium catalysts were prepared from the above mesoporous alumina using tetraamine palladium(II) nitrate solution (Johnson Matthey assay 4.16 % Pd) by the incipient wetness technique. The Pd loading was varied between 0.03 and 4.7 wt%. Dried samples were calcined at 500 °C in static air for 2 h, then reduced under flowing H_2 (10 cm^3.min^{-1}) at 400 °C for 2 h. The resulting samples are termed Pd/meso-Al_2O_3. Amorphous alumina supported catalysts were prepared by the same method using a commercially available Degussa C γ-alumina support (surface area 180 m^{2}g^{-1}). All catalysts were stored in air and not re-activated prior to use.

Characterisation

Bulk Pd concentrations were determined by elemental analysis (AA). Porosity and surface areas were determined on a Micromeritics ASAP 2010 instrument by N_2 adsorption. Surface areas were calculated using the BET equation over the pressure range P/P_0 = 0.02–0.2, where a linear relationship was maintained, while pore-size distributions were calculated using the BJH model up to P/P_0 = 0.6. Surface areas of processed catalysts were 300 ± 20 m^{2}g^{-1}. Metal dispersions were measured using a Micromeritics PulseChemisorb 2700 surface area analyzer by H_2 chemisorption on samples pre-reduced at 400 °C under flowing H_2 for 1 h.

Powder X-ray diffraction (XRD) patterns were recorded using a Bruker AXS D8 diffractometer with a Cu K_α X-ray source. Data were collected in theta-two theta mode over a 2θ range of 10 – 70 ° with a step size of 0.02 ° and dwell time of 50 s per point. Low angle patterns were recorded between 2θ = 0.1 - 5°, with a step size of 0.01 ° and dwell time of 2 s per point.

X-ray photoelectron spectra were acquired at normal emission on a Kratos AXIS HSi spectrometer equipped with a charge neutralizer and Mg K_α excitation source (1253.6 eV). Binding energies were referenced to the adventitious carbon peak (285 eV) and valence band. Wide scans were recorded for surface elemental analysis (pass energy 160 eV), with high resolution spectra recorded at 40 eV pass energy. For the lowest loading sample, 50 scans were necessary to obtain good signal:noise statistics. Spectral fitting was performed using CasaXPS Version 2.3.5 with a common lineshape adopted for all Pd components.

Pd K-edge EXAFS measurements were made on Station 9.3 of the Daresbury SRS facility, using a Si(220) double-crystal monochromator with a beam current/energy of 150 mA / 2 GeV. Transmission

Pd (24.35 keV) K-edge spectra were acquired using powdered catalysts mounted in a stainless steel cell. Dilute samples (≤1 wt% Pd) were measured in fluorescence mode using a 9-channel Ortec ctrain germanium solid-state detector. Spectra were fitted using the Daresbury EXSPLINE and EXCURV98 packages for background subtraction, and phaseshift determination and fitting respectively. Reference transmission spectra of PdO powder (>99 % Lancaster) and a 10 μm Pd foil standard were also recorded. XANES fitting was undertaken using the ATHENA 0.8.050 application from the IFEFFIT Open Source software suite.

High angle annular dark field (HAADF) images were acquired at the Daresbury SuperSTEM 1 facility using a scanning transmission electron microscope (STEM). This instrument is based on a 100 keV VG HB501 with a cold-field emission source, a high-resolution pole piece and Nion Mark II spherical aberration corrector delivering a spatial resolution of 0.1nm. The HAADF image is formed by collecting transmitted electrons on a detector with an angular range from 70 to 210 mrad. These images are sensitive to atomic number and are often referred to as Z-contrast images.

Reactivity Measurements

Alcohol oxidation was performed on a Radleys Carousel reaction station on a 10 ml scale at 60 °C. Reactors were charged with 0.84 mmol of benzyl, cinnamyl and crotyl alcohols (Aldrich 99 %) in toluene, with mesitylene as an internal standard. Reactions were performed under air with 50 mg catalyst and analysed by off-line GC using a DB5 capillary column. Cinnamaldehyde, crotonaldehyde and benzaldehyde were the principal reaction products (87-94 % selectivity) in all cases. There was no evidence for solvent oxidation or alcohol isomerization/polymerization reflecting the mild operating conditions. Reactions were run for up to 24 h with initial rates determined from the linear portion of the reaction profile. Catalyst selectivity and mass balances (closure was >98 %) were determined using reactant and product response factors. Quoted conversions and selectivities are ±2 and ±3% respectively. Recycle tests involved simple catalyst filtration, washing with toluene and air drying.

Results and Discussion

Bulk Structure

The characteristics of the mesoporous alumina support were first explored by N_2 porosimetry (**Figure 1a**) and low angle powder XRD to ensure that Pd impregnation did not result in significant loss of surface area or pore collapse. These revealed the Pd-doped mesoporous alumina retained a high surface area of ~350 m^2g^{-1}, with a well-defined pore network with average pore diameters ~3 nm, and pore wall thicknesses of 1 nm. The hexagonal nature of these pores was confirmed by high resolution STEM (shown later), in accordance with the structure of the template liquid mesophase. The formation of Pd aggregates was also examined by wide angle XRD (**Figure 1b**), and showed the presence of small Pd crystallites between 2-5 nm for metal loadings above 2.4 wt%. Smaller palladium nanocrystals around 1 nm were detected for loadings between 0.5 and 1 wt %, however no Pd reflections were detectable at lower concentrations, consistent with highly dispersed clusters.

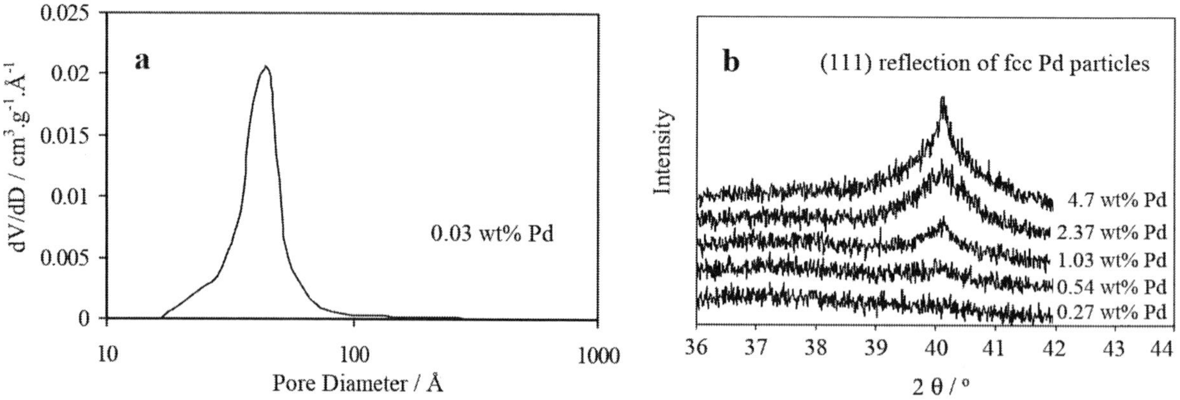

Figure 1. (a) Pore-size distribution of 0.03 wt% Pd/meso-Al_2O_3 sample and (b) powder XRD patterns across Pd/meso-Al_2O_3 series.

Surface Structure

The palladium dispersion and surface oxidation state were subsequently explored by H_2 chemisorption and X-ray photoelectron spectroscopy (XPS) respectively. It should be noted that since the chemisorption method (which titrates metallic sites), requires a pre-reduction step that could in itself induce Pd sintering, the resulting particle size estimates represent an upper limit of the actual values. The resulting particle size estimates ranged between ~0.5 to 5 nm with increasing palladium loading, consistent with the corresponding values determined by XRD. Representative Pd 3d XP spectra from the Pd/meso-Al_2O_3 series are shown in **Figure 2**. The two peaks arise from spin-orbit splitting of the Pd 3d core level, and their binding energy is indicative of the electronic environment of

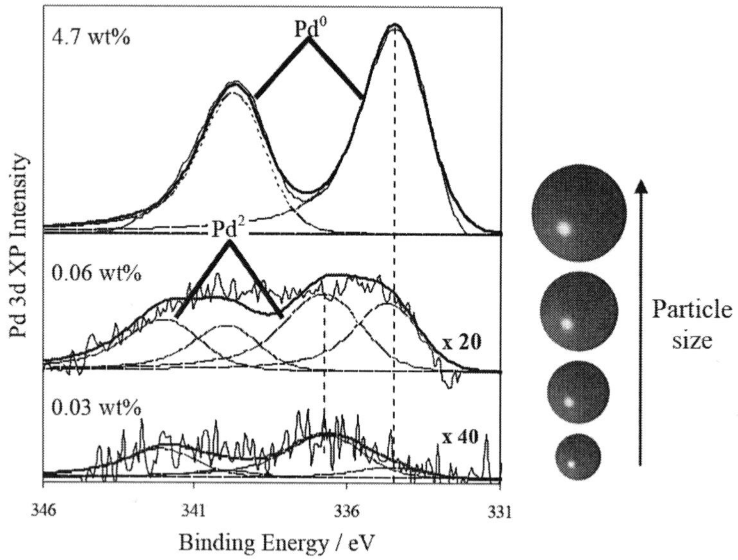

Figure 2. Pd 3d XP spectra of Pd/meso-Al_2O_3 series.

atoms in the surface of supported Pd clusters. These spectra show the expected rise in surface palladium coverage with increasing bulk Pd loading, but also reveal an evolution from a predominantly high binding energy species (336.6 eV) at loadings below 0.5 wt %, to a single, low binding energy species (334.2 eV) at high palladium coverages. Fingerprinting these binding energies against palladium standards suggests a transition from a highly-dispersed electron-deficient surface species, possibly PdO, to metallic Pd nanoparticles with increasing loading. This observation is confirmed by the X-ray absorption (XAS) measurements discussed later, and consistent with the established stabilities of late transition metals versus their metal oxide counterparts, wherein the lower surface energies of the latter favour the formation of oxide terminated surfaces with decreasing particle size.

Catalytic Reactivity

The catalytic performance of these Pd/meso-Al_2O_3 materials was tested in the aerobic oxidation of representative linear and aryl allylic alcohols, namely crotyl and cinnamyl alcohol, and also benzyl alcohol to assess the impact of the C=C group and catalyst versatility towards diverse substrates. Exemplar reaction profiles for the oxidation of crotyl alcohol to crotonaldehyde in a stirred batch system are shown in **Figure 3**. These reveal similar kinetic behaviour for all Pd loadings, with a fast initial turnover followed by a slow deactivation stage. Surprisingly the activities and limiting conversions do not scale with the total palladium loading, but show a pronounced enhancement for more dilute systems.

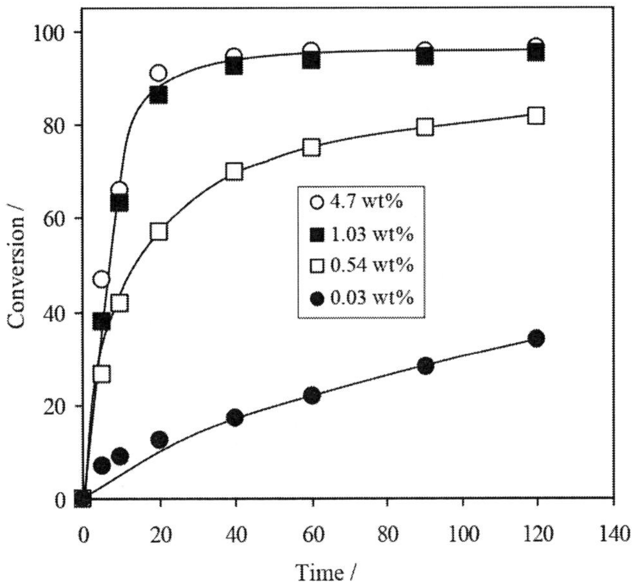

Figure 3. Reaction profiles for the aerobic selective oxidation of crotyl alcohol over representative Pd/meso-alumina samples at 100 °C.

Operando X-ray Absorption Spectroscopy

Further insight into the nature of the palladium species present within both amorphous and meso-tailored alumina support phases was derived from dynamic, real-time studies of the local Pd chemical environment by operando X-ray absorption spectroscopy (XAS).[13a] **Figure 4** shows the resulting evolution in the extended x-ray absorption fine structure (EXAFS) and corresponding white-line intensity in the near-edge region (XANES) during the oxidation of crotyl alcohol. These measurements were made using a purpose-built cell.

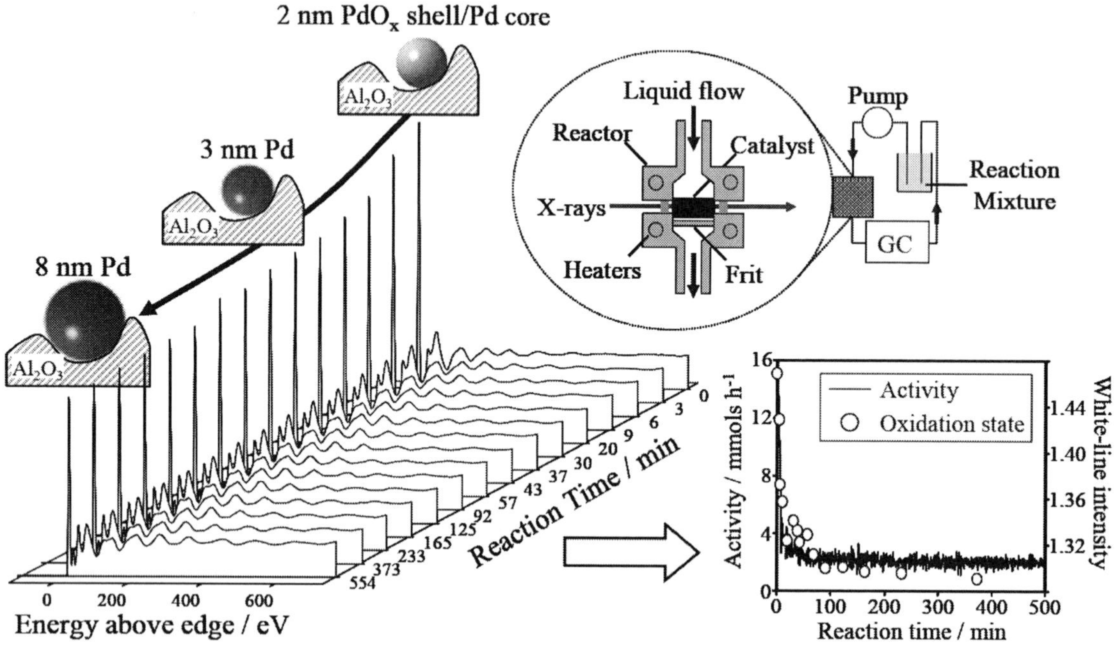

Figure 4. Operando XAS of a 5 wt% Pd/amorphous-alumina sample during the aerobic selective oxidation of crotyl alcohol at 60 °C.

XANES analysis reveals the rapid reduction in the average Pd oxidation state as the reaction proceeds, from partially oxidised 2 nm particles to small entirely metallic particles, and is accompanied by a simultaneous loss in catalyst activity. This fast chemical reduction is followed by a much slower sintering to yield large metallic aggregates, which has negligible impact on the final slow reaction rate. These results provide very strong evidence that under ambient pressure (oxygen-poor) conditions, the active site is palladium oxide, and is supported by pretreatment experiments which show that catalyst activity can be regulated by H_2 or O_2 cycles, which respectively poison or restore selox turnover.

Exchanging the amorphous support for mesoporous support used in the preceding sections reveals a striking difference in the respective stabilities of the surface Pd species under reaction conditions. In contrast with the amorphous support, extremely dilute loadings resulted in entirely oxidic palladium, with a small first coordination shell indicative of a truncated PdO_x environment with no evidence for any Pd metal. The first shell coordination number was indicative of sub-nanometre Pd clusters, in accordance with the powder XRD and dispersion measurements. These electron-deficient Pd species were completely stable under reaction conditions (**Figure 5**), showing neither reduction nor agglomeration over 24 h on-stream.

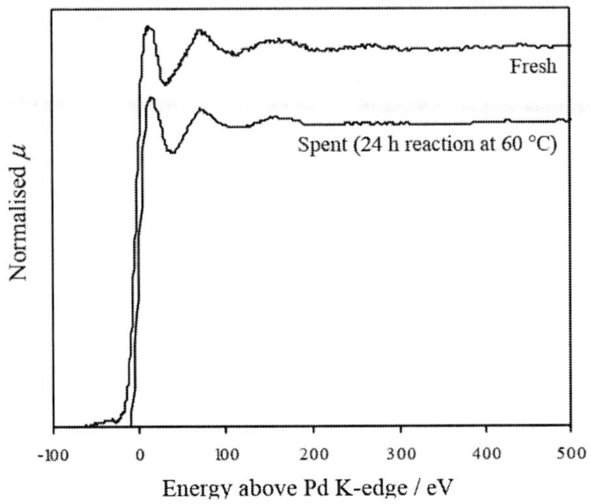

Figure 5. Operando XAS of a 0.03 wt% Pd/meso-alumina sample during the aerobic selective oxidation of crotyl alcohol at 60 °C.

Transmission Electron Microscopy

Further insight into the nature of these highly stable electron-deficient Pd clusters was obtained from scanning transmission electron microscopy (STEM).[15] **Figure 6** shows two high angle annular dark field STEM images of the fresh 0.03 wt% Pd/meso-alumina sample, this mode is sometimes called Z-contrast imaging, and is especially sensitive to heavy atoms. Both images clearly show the hexagonal mesopores of the alumina support phase, and identify the active palladium species as individual, atomically-dispersed atoms, most likely embedded into the surfaces of the alumina framework.

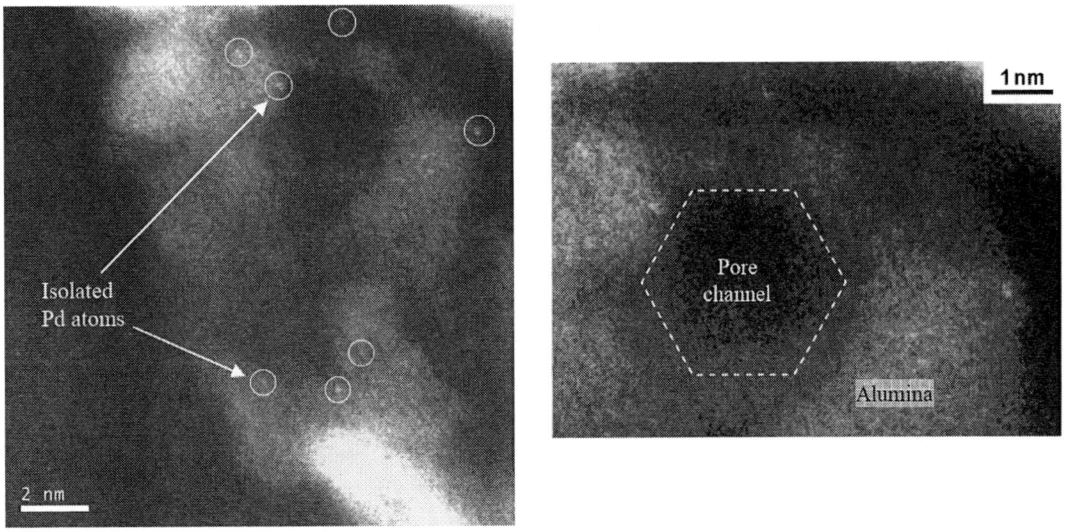

Figure 6. HAADF-STEM images of a 0.03 wt% Pd/meso-alumina sample visualising the alumina support mesopores and isolated Pd^{x+} atoms.

The density of such isolated Pd^{2+} atoms over the mesoporous alumina correlates directly with the resultant catalyst turnover frequency in crotyl and cinnamyl alcohol selective oxidation (**Figure 7**), confirming our hypothesis that oxide-like palladium centres are active sites for these reactions. Indeed the activity for catalysts derived using amorphous alumina exhibit the same general trends, suggesting large metal clusters are intrinsically less active than either surface palladium oxide or isolated electron-deficient Pd atoms. This observation is consistent with our recent time-resolved XPS studies over Pd(111) model surfaces,[16] which show that metallic palladium favours decarbonylation of the desired aldehyde product and associated self-poisoning: surface oxygen suppresses this pathway. It should also be noted that the absolute turnover frequencies are the highest reported to date for aerobic for crotyl alcohol oxidation over any heterogeneous catalyst.

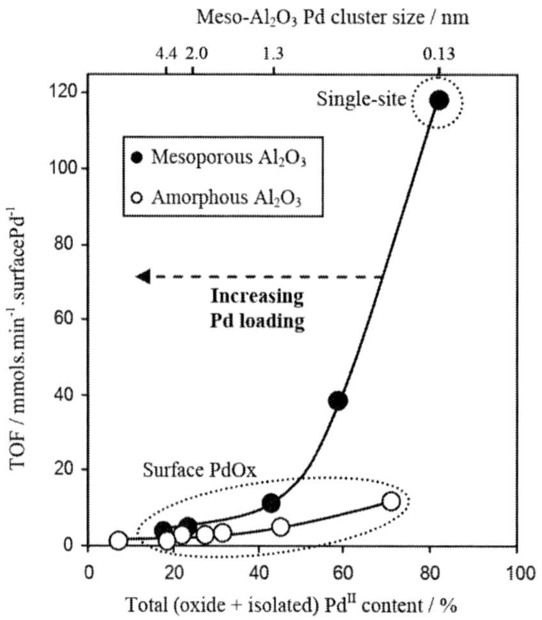

Figure 7. Structure-reactivity correlations in crotyl alcohol selox over Pd/alumina catalysts.

Conclusions

Over a low surface area alumina support, low palladium concentrations stabilise small Pd particles, comprising a metallic core and capping oxide surface layer. Use of a high area, mesoporous alumina phase dramatically increases the palladium dispersion, favouring isolated atoms which are all electron-deficient. In both systems the allylic alcohol selox is strongly favoured over electron-deficient palladium species.

References

1. Mallat, T.; Baiker, A. *Chem. Rev.* **2004**, *104*, 3037-3058.
2. Sheldon, R.A. *Pure Appl. Chem.* **2000**, *72*, 1233-1246.

3. Mallat, T.; Baiker, A. *Catal. Today* **1994**, *19*, 247-284.
4. Hardacre C.; Mullan, E.A.; Rooney, D.W., Thompson, J.M. *J. Catal.* **2005**, *232*, 355-365.
5. Kimmerle, B; Grunwaldt J.-D.; Baiker, A. *Top. Catal.* **2007**, *44*, 285-292.
6. Mallat, T.; Bodnar, Z.; Hug, P.; Baiker, A *J. Catal.* **1995**, *153*, 131-143.
7. Yamaguchi, K,; Mizuno, N. *Angew. Chem. Int. Ed.* **2002**, *41*, 4538-4542
8. Enache, D.I.; Edwards, J.K.; Landon, P.; Solsona-Espriu, B.; Carley, A.F.; Herzing, A.A.; Watanabe, M.; Kiely, C. J.; Knight, D.W.; Hutchings, G. J. *Science* **2006**, *311*, 362-365.
9. Hughes, M.D.; Xu1, Y.-J.; Jenkins, P.; McMorn, P.; Landon, P.; Enache, D.I.; Carley, A.F.; Attard, G.A.; Hutchings, G.J.; King, F.; Stitt, E.H.; Johnston, P.; Griffin, K.; Kiely, C.J. *Nature* **2005**, *437*, 1132-1135.
10. Alardin, F.; Delmon, B.; Ruiz, P.; Devillers, M. *Catal. Today* **2000**, *61*, 255.
11. b) Grunwaldt, J.-D.; Caravati, M.; Baiker, A. *J. Phys. Chem. B* **2006**, *110*, 25586- 25589.
12. Corma, A. *Chem. Rev.* **1997**, *97*, 2373-2419.
13. a) A. F. Lee, K. Wilson, *Green Chem.* **2004**, *6*, 37-42; b) A. F. Lee, S. F. J. Hackett, J. S. J. Hargreaves, K. Wilson, *Green Chem.* **2006**, *8*, 549-555; c) J. D. Grunwaldt, M. Caravati, M. Ramin, A. Baiker, *Catal. Lett.* **2003**, *90*, 221-229.
14. Vaudry, F.; Khodabandeh, S.; Davis, M.E. *Chem. Mater.* **1996**, *8*, 1451-1464.
15. Hackett, S.F.J.; Brydson, R.M.; Gass, M.H.; Harvey, I.; Newman, A.D.; Wilson, K.; Lee, A.F. *Angew. Chem. Int. Ed.* **2007**, *46*, 8593-8596.
16. Lee, A.F.; Chang, Z.; Ellis, P.J.; Hackett, S.F.J.; Wilson, K. *J. Phys. Chem. C* **2007**, *111*, 18844-8847.

Synthesis of Improved Fuel Cell Catalysts Using Electroless Deposition Methods

K. Beard, J.W. Van Zee, J.R. Monnier

Introduction

In an effort to reduce the amount of noble precious metals used in electrochemical catalysts and to increase catalytic activity, a novel method of synthesis that involves electroless deposition (ED) of a Pt salt on a carbon support that has been previously seeded with a group VIIIB metal has been developed. Our previous work [1,2] has focused on the ED of Pt onto other noble metals such as Rh and Pd. This methodology has been extended to include deposition of Pt on base metals as well as deposition of base metals on Pt. Others [1] have shown that electronic interactions between Pt and base metals such as Co, Ni, and Cu may occur, which can increase the activity of the Pt towards oxygen reduction. However, the highly acidic environment of a fuel cell typically corrodes the base metal, leading to reduced performance. Thus, a core-shell structure of a Pt shell over a base metal core would be advantageous to gain the activity enhancements of the Pt-base metal interaction while mitigating the corrosion of the base metal, now protected by the Pt shell. This presentation illustrates aspects of ED to give this preferred core-shell type structure as well as how the controlled deposition of Pt gives an overall reduction in metal particle size and a more efficient utilization of the expensive Pt component. X-ray photoelectron spectroscopy (XPS) is used to provide strong evidence of a core/shell and solid solution for Pt electrolessly deposited on Co and Pd, respectively, while HRTEM confirms the synthesis of smaller Pt-containing particles using ED methods. The oxygen reduction activity of the synthesized ED catalysts is determined using a rotating disc electrode (RDE) to correlate the promoter effect of possible electronic interactions between the precursor base metal and the electrolessly-deposited Pt on overall fuel cell performance.

Materials and Methods

The preparative method consists of using commercial carbon black which has been seeded with a variety of Group VIIIB metals using conventional, wet impregnation techniques. Platinum metal deposition from reduction of $PtCl_6^{2-}$ ions occurred on the seeded metal particles by the Group VIIIB metal-catalyzed activation of a reducing agent, dimethylamine borane (DMAB) in our studies, at the seed sites. Therefore, the catalytically active metal site serves as both anode and cathode for the reduction of $PtCl_6^{2-}$ ions. This methodology should result in a core/shell geometry for the catalyst particles, assuming the deposited Pt metal is itself catalytic (autocatalysis) for further electroless deposition. Deposition does not occur on the carbon support since there are no sites capable of activating the reducing agent. The kinetic dependencies of the Pt deposition on metal surface include pH of the ED solution as well as concentrations of reducing agent, Pt^{4+} salt source, and Pt^{4+} complexing agent.

Prepared catalysts were analyzed using AA, TEM and HRTEM, selective H_2 chemisorption, oxygen reduction with rotating disk electrode (RDE), and XPS. AA was used to determine final weight loadings of both platinum and the Group VIIIB seed metal. Particle sizes and metal dispersions were determined for the Pt and the Group VIIIB metal using, TEM, HRTEM, and H_2 chemisorption. Electrochemically-active surface area and oxygen reduction activity were established by RDE studies. Lastly, verification of core/shell structures, surface compositions, and evidence of electronic interactions between Pt and the precursor metal were established with XPS.

Results and Discussion

TEM Analysis

TEM characterization has shown that ED methods form Pt particles smaller than those present in a commercially-available 20% Pt/C electrocatalyst to give 30% more Pt surface sites than the commercial 20% Pt/C sample. Further, the ED-derived samples also have considerably narrower particle size distributions, as seen in Figure 1. The overall smaller Pt particles as well as the narrower particle size distributions provide increased dispersion of Pt. Table I tabulates the important data taken from Figure 1 which shows that the number of Pt particles/g_{cat} for the ED catalysts and the standard sample are quite similar, despite the ED samples containing only 40% of the total Pt in the standard 20% Pt/C sample. Also, the concentration of Pt surface sites for the best ED catalysts is approximately 60% that of the standard catalyst while containing approximately 60% less Pt. Thus, the concentration of surface Pt sites and Pt particles, once normalized to Pt loading, are greater for the ED catalysts compared to the standard catalyst.

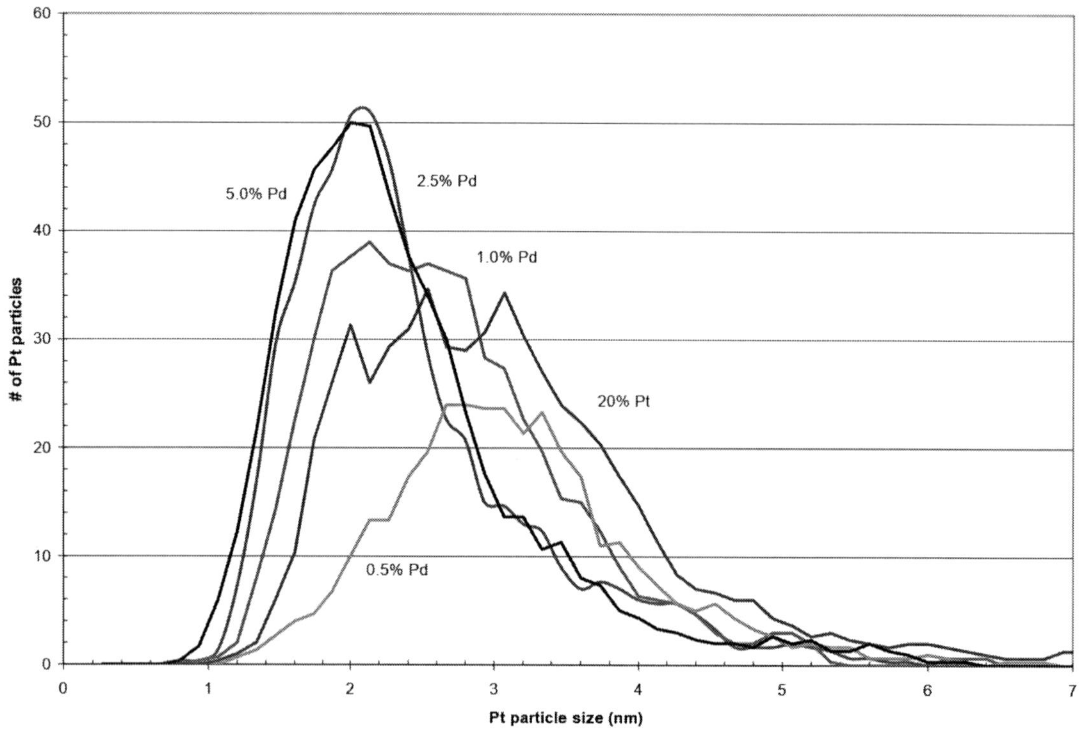

Figure 1: Particle size distribution curves for the ED catalysts (8.0% Pt on 5.0%, 2.5%, 1.0%, and 0.5% Pd seeds on carbon) and the standard catalysts (20% Pt on carbon).

Table I: Tabulated data from analysis of TEM images and Figure 1.

		TEM Results			
Pd Loading	Pt Loading	Avg Particle Diameter	Dispersion	Concentration of surface Pt sites/g_{cat}	Concentration of Pt particles
%	%	Å	%		particles/g_{cat}
n/a	19.2	40	28	1.7E+20	4.4E+17
0.5	8.0	38	32	7.0E+19	1.8E+17
1.0	8.0	32	36	9.2E+19	3.3E+17
2.5	8.0	30	38	9.8E+19	4.2E+17
5.0	8.0	32	36	9.3E+19	4.1E+17

Determining Particle Structure with XPS

The structure of the catalyst particles synthesized by ED was examined by XPS. The Pd 3d orbitals were analyzed by XPS for 0.5%, 1.0%, 2.5%, and 5.0% Pd/C samples and 7% Pt-0.5% Pd/C, 8% Pt-1.0% Pd/C, 8% Pt-2.5% Pd/C, 8% Pt-5.0%Pd/C samples; thus, nearly identical quantities of Pt have been electrolessly deposited on varying quantities of Pd. As shown previously in Figure 1, the average Pt particle size decreases with increasing Pd loading. Another factor to consider here is the escape depth of photoelectrons. Seah and Dench [2] estimated the photoelectron escape depth to be approximately 6 to 7 monolayers (corresponding to 18 to 21 angstroms) through an element at a binding energy of approximately 330 eV. Thus, the XPS is not a strictly surface-sensitive instrument and is instead a near-surface experiment.

It was originally surmised that the Pt-Pd particles would have a core/shell structure because Pt must be initially deposited on top of the Pd precursor catalyst during the process of electroless deposition. Furthermore, the reducing agent DMAB kinetically favors deposition of Pt on Pd rather than Pt on Pt; thus, the Pd should be first covered with Pt before autocatalytic deposition of Pt occurs. Figure 2 shows an example of some early data which supported this first hypothesis. The plots show the change in the Pd 3d orbital as analyzed by XPS for 0.5%, 1.0%, 2.5%, and 5.0% Pd/C before and after approximately 8% Pt is electrolessly deposited on the Pd precursor. In the case of 0.5% and 1.0% Pd/C, a strong (although not complete) suppression of the Pd 3d orbital is observed which is not true for the 2.5% and 5.0% Pd/C samples. Deconvolution of the Pd 3d peaks after electroless deposition of Pt, indicated that every peak contained both Pd^0 and Pd^{2+} states. These results shed doubt on the core/shell structure hypothesis because, if fully encapsulated, the Pd should remain in a metallic (Pd^0) state since it should be protected from the oxidizing environment (room temperature air) by the Pt overlayer.

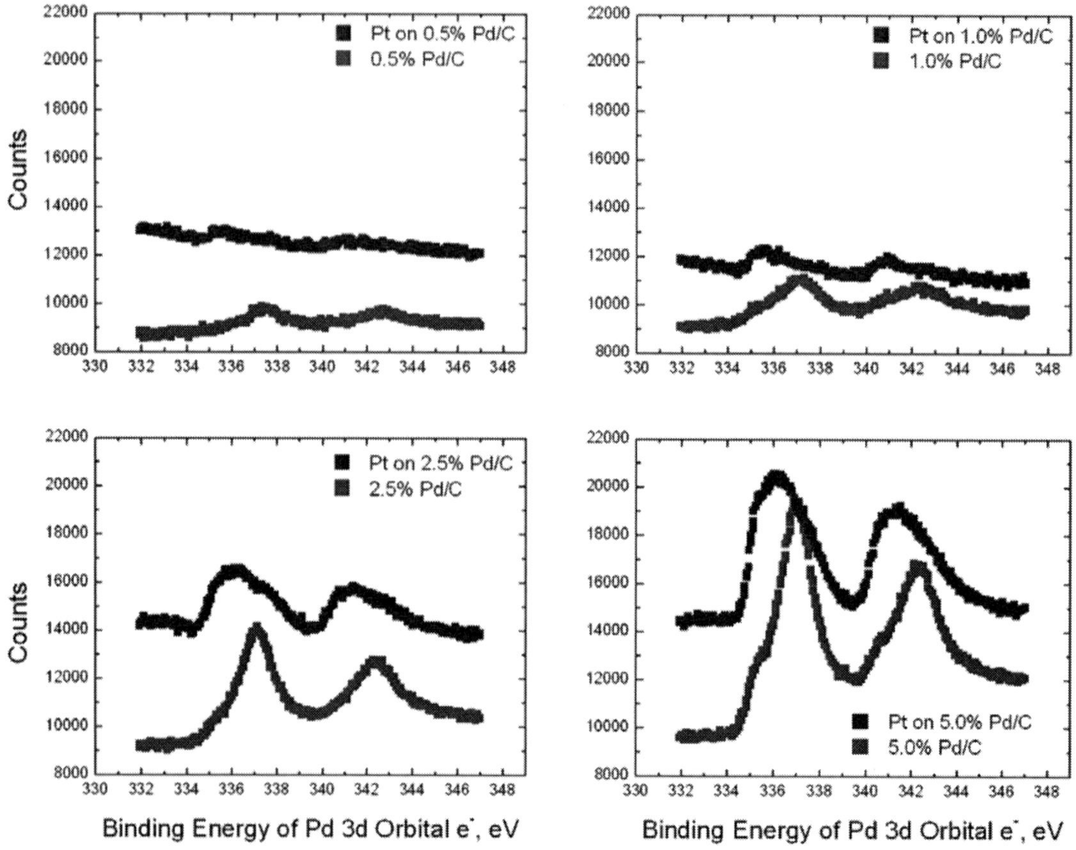

Figure 2: Pd XPS spectra for similar weight loadings of Pt deposited on varying amounts of Pd.

The results of deconvolution are presented below in Table II as the fraction of Pd which is in the metallic state. This fraction is very low prior to electroless deposition of Pt; however, the fraction only reaches approximately 60% even after the Pt is deposited. It is also interesting to note that the fraction of Pd in the metallic state does not change appreciably with Pd loading which dictates the relative amount of Pt in each Pt-Pd particle.

Table II: Fraction of metallic Pd at near surface of Pt-Pd nanoparticles

Sample	$\frac{Pd^0}{Pd^0+Pd^{2+}}$ Pd3d$_{5/2}$	$\frac{Pd^0}{Pd^0+Pd^{2+}}$ Pd3d$_{3/2}$
0.5% Pd/C	0.06	0.06
1.0% Pd/C	0.15	0.15
2.5% Pd/C	0.08	0.08
5.0% Pd/C	0.16	0.17
7% Pt on 0.5% Pd/C	0.55	0.56
8% Pt on 1.0% Pd/C	0.55	0.56
8% Pt on 2.5% Pd/C	0.57	0.57
8% Pt on 5.0% Pd/C	0.46	0.47

Another series of experiments using in-situ pretreatments was carried out. For this experiment, all bimetallic compositions were reduced at 200°C in H_2; however, Pt on 0.5% Pd/C and 1.0% Pd/C were also reduced at a lower temperature (75°C) before reduction at 200°C to ensure that any change in composition was the result of reduction only and not intraparticle diffusion. Further, these two samples were subjected to a heat treatment in vacuum at 400°C (Pt on 0.5% Pd/C), 540°C, and cooling from 540°C to room temperature (Pt on 1.0% Pd/C) after reduction. The results of these experiments are summarized in Table III. All samples showed some increase (approximately 10 - 20%) in surface molar concentration of Pd after the initial reduction. Subsequent treatments (higher reduction temperature, heat treatment, cooling) had only marginal effects on the surface composition. Finally, the observed Pd surface concentration was always less than the bulk concentration of Pd (aside from Pt on 5.0% Pd/C after reduction) indicating a Pt-rich near surface region. The composition of this near surface region is clearly proportional to bulk composition (e.g. increasing Pd loading from 0.5% to 1.0%

approximately doubles the near-surface concentration of Pd). However, the apparent equilibrium reached after the first reduction suggests the presence of a Pt-rich solid solution in the near surface area of the nanoparticles which is unaffected by heat treatment.

Table III: Effect of Reduction and Heat Treatment on Near-Surface Composition

Sample	Treatment	Mole Conc. Pd (%)	Bulk Mole Pd (%)	Mole Conc. Pt (%)	Bulk Mole Pt (%)
Pt on 0.5% Pd/C	As Prep	6.8	11.6	93.2	88.4
	H_2 75^{0}C	7.6	11.6	92.4	88.4
	H_2 200^{0}C	7.9	11.6	92.1	88.4
	Vac 400^{0}C	8.6	11.6	91.5	88.4
	Vac 540^{0}C	7.2	11.6	92.8	88.4
Pt on 1.0% Pd/C	As Prep	10.6	18.6	89.4	81.4
	H_2 75^{0}C	12.9	18.6	87.1	81.4
	H_2 200^{0}C	15.5	18.6	84.5	81.4
	Vac 540^{0}C	14.6	18.6	85.4	81.4
	Cool to RT	16.7	18.6	83.3	81.4
Pt on 2.5% Pd/C	As Prep	27.0	36.4	93.2	63.6
	H_2 200^{0}C	33.2	36.4	66.8	63.6
Pt on 5.0% Pd/C	As Prep	49.8	53.4	50.2	46.6
	H_2 200^{0}C	53.5	53.4	46.2	46.6

References

1. T. Toda, H. Igarashi, H. Uchida, M. Watanabe, *Journal of the Electrochemical Society*, 146 (1999) 3750 - 3756.
2. M.P.~Seah, and W.A.~Dench, Surface and Interface Analysis, 1 (1979) 2-11.

Simulation of Heterogeneously Cs-Doped Heteropolyacid Catalyzed Transesterification for Biodiesel Production

Biodiesel is a promising nontoxic and biodegradable fuel comprised of alkyl esters of long chain fatty acids, which are derived from renewable biological sources such as vegetable oils and animal fats[1]. It has similar physical properties to petrochemical diesel and is considered the best alternative fuel candidate for use in diesel engines[2]. Biodiesel was firstly tested as engine fuel by Rudolf Diesel one hundred years ago[3]. However, it was only used in emergency situations in past few decades. Recently, due to the increases in crude oil prices, limited resources of fossil oil and the environmental concerns, biodiesel is becoming a developing area of high concern. Compared to diesel fuel, one of the advantages of biodiesel is that the raw materials used to produce it are natural and renewable, and therefore biodiesel has low emission profiles, less toxic, and is more environmentally beneficial.

Different studies have been carried out extensively using different ways of biodiesel production, with different oils as raw materials[4]. The most commonly used method is by transesterification of triglycerides, the main constituent of vegetable oils and animal fats, with an alcohol to form esters and glycerol. Transesterification (also called alcoholysis) is the reaction of a fat or oil with an alcohol to form esters and glycerol. Methanol and ethanol are most frequently used alcohols in the transesterification processes, especially methanol, because of its low cost and its physical and chemical advantages[5]. Because the reaction is reversible, excess alcohol is used to shift the equilibrium to the products side. Different kinds of catalysts, homogeneous ones such as sodium hydroxide, potassium hydroxide, sulfuric acid and supercritical fluids, and heterogeneous ones such as lipases, can be used to improve the reaction rate and yield. Currently, homogeneous alkaline catalysts, such as *K* or *Na* alkoxides or hydroxides are most often used commercially for the transesterification of biodiesel production[6]. However, removal of the soluble base after reaction is a major problem, because aqueous quenching results in the formation of stable emulsions and saponification, rendering separation and purification of the methyl ester difficult. As a result, biodiesel production by these routes is still not cost-competitive with petrochemical diesel fuel[7]. These disadvantages could, however, be removed by use of a heterogeneous catalyst, which can eliminate of a quenching step (and associated contaminated water waste) to isolate the products and operate a continuous process[8]. Heterogeneous catalysts, including zeolites[9], alkali earth oxides[10] and hydrotalcites[11,12], have been investigated in transesterification reactions.

Heteropolyacids (HPAs) are an interesting class of solid acid catalysts, which can be used for diverse reactions that require strong acidity. Although the acid forms are themselves useful solid catalysts, they are highly soluble in, and difficult to separate from polar media. This high solubility makes such solid acid catalysts unsuitable for catalyzing polar chemistry[13]. Recently, Narasimharoa et al.[6] introduced a series of

Cs-doped HPAs (e.g., $Cs_xH_{3-x}PW_{12}O_{40}(x=1-3)$), which exhibit dramatic increases in surface area and profound changes in solubility over the parent HPA. Partial substitution of protons by Cs^+ also changes the number of available surface acid sites with significantly higher activity than the parent acid[14]. In their study[6], this series of insoluble HPA salts of general formula $Cs_xH_{3-x}PW_{12}O_{40}(x=1-3)$ was characterized using a range of bulk and surface sensitive probes, and investigated for solid acid transesterification reactions. However, few studies concerning the kinetic mechanism of this Cs-doped heteropolyacid catalyzed transesterification have been proposed. Thus, in our recent work, a heuristic diffusion-reaction kinetic model has been established to simulate the transesterification of alkyl ester with methanol over a series of heterogeneous Cs-doped heteropolyacid catalysts.

Firstly, we applied the Eley-Rideal type of mechanism[15] to the simulation of the heterogeneous transesterification of biodiesel production over Cs-doped heteropolyacid catalysts. The transesterification of tributyrin by methanol consists of the following three consecutive transesterification reactions:

$$T + MeOH \Leftrightarrow D + MeBu$$

$$D + MeOH \Leftrightarrow M + MeBu$$

$$M + MeOH \Leftrightarrow G + MeBu$$

The triglyceride, tributyrin (T), firstly reacts with methanol, to produce a diglyceride, dibutyrin(D), and the fatty acid ester, in this case methylbutanoate (MeBu).Then the diglyceride formed reacted further with the alcohol to form another molecule of the fatty acid ester and a monoglyceride, monobutyrin (M). This then reacts with the alcohol to obtain glycerine (G) and the fatty acid ester.

The elementary steps and the corresponding reaction mechanism for each transesterification reaction are shown in Table 1. The three parameters of the model are the activation energy $E_{a,MeOH}$, the pre-exponential factor A of the rate-determining step, and the adsorption equilibrium coefficient K_A. The rate equations for the three glyceride transesterification are presented as follows

$$r_1 = \frac{k_{MeOH}(a_{MeOH} - (1/K_{eq})(a_{MeBu}a_D / a_T))}{1 + (K_A / K_{eq})(a_{MeBu}a_D / a_T) + K_A a_D + K_A a_M + K_A a_G} \quad (1)$$

$$r_2 = \frac{k_{MeOH}(a_{MeOH} - (1/K_{eq})(a_{MeBu}a_M / a_D))}{1 + (K_A / K_{eq})(a_{MeBu}a_M / a_D) + K_A a_D + K_A a_M + K_A a_G} \quad (2)$$

$$r_3 = \frac{k_{MeOH}(a_{MeOH} - (1/K_{eq})(a_{MeBu}a_G / a_M))}{1 + (K_A / K_{eq})(a_{MeBu}a_G / a_M) + K_A a_D + K_A a_M + K_A a_G} \quad (3)$$

And

$$E_{a,MeOH} = -RTLn(\frac{k_{MeOH}}{A}) \quad (4)$$

Where k_{MeOH} is the methanol adsorption rate constant equal for all three reactions, and a_i is the activity of each reaction component i. K_{eq} is the equilibrium coefficient of the overall transesterification reaction.

Table 1 Elementary steps and reaction mechanism for the biodiesel transesterification reactions

Elementary reactions	Reaction mechanism		
	T-D	D-M	M-G
$CH_3OH + * \Leftrightarrow CH_3OH *$	1	1	1
$CH_3OH * + T \Leftrightarrow D * + MeBu$	1		
$CH_3OH * + D \Leftrightarrow M * + MeBu$		1	
$CH_3OH * + M \Leftrightarrow G * + MeBu$			1
$D* \Leftrightarrow D + *$	1		
$M* \Leftrightarrow M + *$		1	
$G* \Leftrightarrow G + *$			1

Since it was observed that the reaction mixture was not ideal, activities, rather than concentrations should be used in the rate expression. The activity a_i for each component i is calculated from the component concentration C_i and the activity coefficient γ_i (Eq. (5)). The UNIFAC contribution method was used to calculate activity coefficients [16].

$$a_i = C_i \times \gamma_i \tag{5}$$

Thus for a batch reactor with the assumption that no by-product is observed, differential equations of reaction components are expressed as:

$$\frac{dn_T}{dt} = -r_1 W \tag{6}$$

$$\frac{dn_D}{dt} = r_1 W - r_2 W \tag{7}$$

$$\frac{dn_M}{dt} = r_2 W - r_3 W \tag{8}$$

Where n_T is the number of mole of triolein, n_D is the number of mole of diolein, and n_M is the number of mole of monoolein. W is the mass of the catalyst and r_1, r_2, r_3 are the reaction rate derived from Eqs. (1)-(3). And the estimation of the kinetic parameters was performed by minimization of the objective function S:

$$S=\sum_{k=1}^{N}(n_{T,k}-\hat{n}_{T,k})^2+\sum_{k=1}^{N}(n_{D,k}-\hat{n}_{D,k})^2+\sum_{k=1}^{N}(n_{M,k}-\hat{n}_{M,k})^2 \qquad (9)$$

Where $\hat{n}_T$, $\hat{n}_D$, $\hat{n}_M$ are the kth experimental number of moles of triglyceride, diglyceride and monoglyceride, and n_T, n_D, n_M are the corresponding value calculated by the model. N is the total number of experimental values. The minimisation is achieved by a Genetic algorithm developed by MATLAB.

For the kinetics model using Eley–Rideal type of mechanism we proposed above, it is assumed that there is no mass transfer limitation at the surface of particles and hence the concentration at the surface of a particle is equal to the bulk concentration. Another important assumption in this model is the presence of high mass transport limitations inside the catalyst particles as a result of the relatively low diffusivity of triglycereide. Thus, no diffusion of reactants inside the pellet has been considered. However, for a given fixed bed reactor, the effect of mass transfer inside the homogeneous catalyst bed should be taken into account[17]. Thus, a diffusion-reaction bulk scale model for a fixed bed batch reactor is developed. The mass balance for each bulk specie i can be expressed as a function of diffusion transfer (effective diffusivity $D_{L,i}$), convection (axial velocity u), and rates of reaction ($r_{cat,j}$) (Eq. (10)).

$$\frac{\partial C_i}{\partial t}=\frac{D_{L,i}}{\varepsilon}\left[\frac{\partial^2(C_i)}{\partial z^2}\right]-\frac{u}{\varepsilon}\frac{\partial C_i}{\partial z}+\frac{(1-\varepsilon)}{\varepsilon}\rho_{cat}\sum_{j=1}^{nrxn}v_{ij}r_j \qquad (10)$$

Where C_i represents the bulk concentration of specie i, z is the bed axial length, and ε is the bed void fraction. ρ_{cat} is the density of catalyst, v_{ij} represents the stoichiometric coefficient for specie i in reaction j, and r_j is the rate of jth reaction (derived from Eqs. (1)-(3)).

The novelty of this framework lies in detailed modeling of surface reacting kinetic phenomena and integrating that with particle-level transport phenomena all the way through to process design and optimisation, which has been done for biodiesel production process for the first time. This multi-disciplinary research offers better insights into catalyst design and process intensification for the industrial application of Cs-doped heteropolyacid catalysts for biodiesel production. A case study of the transesterification of tributyrin with methanol has been demonstrated to establish the effectiveness of this methodology.

References

1. Krawczyk T. Biodiesel - Alternative fuel makes inroads but hurdles remain. INFORM 7, 1996; 801-829.
2. Romano S. Vegetable Oils Fuels, in: Proc. Int. Conf. on Plant and Vegetable Oils as Fuels, ASAE, MI, USA, 1982; p. 106.
3. Shay E. G. Diesel fuel from vegetable oils: status and opportunities. Biomass and Bioenergy. 1993; 4, 227-242.
4. Marchetti J.M., Miguel V.U., and Errazu A.F. Possible methods for biodiesel production. Renewable and Sustainable Energy Reviews. 2005; 6:1300-1311.
5. Ma F., Hanna M. A. Biodiesel production: a review. Biosource technology. 1999; 70:1-15.
6. Narasimharao K., Brown D.R., Lee A.F., Newman A.D., Siril P.F., Tavener S.J., Wilsona K. Structure–activity relations in Cs-doped heteropolyacid catalysts for biodiesel production. Journal of Catalysis. 2007; 248: 226–234.
7. Freedman B., Pryde E.H., Mounts T.L., Am J. Oil Chem. 1984; 61: 1638.
8. Ono Y., Baba T.. Catal. 1997; 38: 321.
9. Leclercq E., Finiels A., Moreau C, Am J.. Oil Chem. 2001; 78: 1161.
10. Gryglewicz S.. Bioresour. Technol. 1999; 70: 249.
11. Cantrell D.G., Gillie L.J., Lee A.F., Wilson K., Appl. Catal. A Gen. 2005; 287: 183
12. Di Serio M., Ledda M., Cozzolino M., Minutillo G., Tesser R., Santacesaria E., Ind. Eng. Chem. Res. 2006; 45: 3009.
13. Okuhara T., Mizuno N., Misono M.. Adv. Catal. 1996;41: 113.
14. Mizuno N., Misono M.. Chem. Lett. 1987; 967.
15. Dossin T.F., Reyniers M. F., Marin G.B.M., Appl. Catal. B Environ. 2006; 61: 35-45.
16. Reid C., Prausnitz J.M., Poling D.E.. The Properties of Gases and Liquids, fourth and fifth ed., McGraw-Hill, New York, 1987.
17. Ding Y.. Alpay E. Adsorption-enhanced steam-methane reforming. Chemical Engineering Science. 2000; 55:3929-3940.

205b Growth of Fe Based Nanocatalysts on Reconstructed Au(111) Surfaces

Christopher Matranga, National Energy Technology Laboratory, P. O. Box 10940, Pittsburgh, PA 15236 and Neetha Khan, Chemistry and Surface Science Division, U. S. Dept. of Energy, 626 Cochrans Mill Road, Pittsburgh, PA 15236.

Iron-based catalysts, including iron oxides, are an important class of materials with relevance to Fischer-Tropsch catalysis and gas-sensing applications. By growing nanostructured particles on single-crystal surfaces, we can create a model system to study size and shape effects on reactivity. We have studied the formation of monolayer thick iron oxide nanoparticles and thin films on the reconstructed Au(111) surface. STM, XPS, ISS, and LEED were used to evaluate the structure and composition of the iron oxide nanoparticles and films as a function of growth conditions. Iron oxide was grown by depositing iron on Au(111), followed by oxidation at room temperature and annealing to 700 K. XPS results indicate that the Fe is oxidized at room temperature, but the STM results indicate that the particles are not ordered until after annealing to 700 K. Atomically-resolved STM images show that at 0.3 ML Fe coverage, iron oxide nanoparticles, pseudo-hexagonal in shape are formed, with large defects occurring in the corners. STM images of FeO particles over 0.5 ML Fe also show evidence of a non-coincidence overlayer lattice with a short periodicity of 0.25-0.3 nm modulated by a larger periodicity of approximately 3.5 nm. The larger periodicity results from a moiré pattern formed between the iron oxide overlayer and the underlying Au(111) surface.

33c Direct Dimethyether Synthesis Using Revese Shift Assisted Reforming

Kwang-Deog Jung[1], Oh-Shim Joo[1], Heon Jung[2], and Dae Ryook Yang[2]. (1) Clean Energy Research Center, Korea Institute of Science and Technology, 39-1 Hawolgok-dong Sungbuk-gu, Seoul, South Korea, (2) Chemical and Biological Engineering, KOREA University, Anam-dong Sungbuk-gu, Korea University, Seoul, South Korea

DME has been attentioned as a fuel in Asia region. Although a direct DME synthesis process from synthesis gas seems to be attractive in the thermodynamic point of view, a lot of CO2 production results in the loss of a overall DME yield in stoichiometric point of view. On the other hand, since the direct DME synthesis can be optimized at about 1 of H2/CO ratio, the refoming of methane has been an important issue to ajust the H2/CO ratio to about 1. In our direct DME process, the conventinal steam reforming is used. The synthesis gas with the excess hydrogen from the steam reformer is separated to adjust the H2/CO ratios of the DME feed gas by a hydrogen membrane. The separated hydrogen is used to convert CO2 to CO on the reverse shift reaction, and then the enriched CO gas is recycled and combined with the synthesis gas from the reformer. CO2 source would be the additional one or the recovered one from the DME reactor. This process can increase the overall DME yield by mitigating CO2. The bench plant of this DME process was intalled and operated to show the high yield and low CO2 production in the direct DME synthesis, as compared with the conventional direct DME process.

172e Performance Comparison of a Bubble Reactor and a Loop Reactor for Preparing a C5 Hydrofluoroether, HFE-449mec-f, in the Gas-Liquid Biphasic Reaction

Hyunjoo Lee[1], Prabal Sapkota[1], Sang Deuk Lee[1], Hoon Sik Kim[2], Kwang Ho Song[3], and **Honggon Kim**[1]. (1) Energy and Environment Research Division, Korea Institute of Science and Technology, 39-1, Hawolgok-dong, Seongbuk-gu, Seoul 136-791, South Korea, (2) Dept. of Chemistry, Kyung Hee University, 1 Hoegidong, Dongdaemoongu, Seoul 130-701, South Korea, (3) Department of Chemical and Biological Engineering, Korea University, Anamdong, Seongbukgu, Seoul, South Korea

Hydrofluoroethers(HFEs) have been suggested as alternatives to chlorofluorocarbons(CFCs), hydrochlorofluorocarbons(HCFCs) and hydrofluorocarbons(HFCs) because of their low global warming potential and zero ozone depletion potential. HFE-449mec-f(C3HF6OC2H2F3), a C5 HFE, is a plausible alternative cleaning solvent which has high dissolving capability on fluoro oils. This study deals with the preparation of HFE-449mec-f in a bubble reactor and a buss loop reactor, both of which efficiently carry out gas-liquid biphasic reactions. HFE-449mec-f can be synthesized by hydroalkoxylation of hexafluoropropene(HFP) in 2,2,2-trifluoroethanol(TFEA) with fluoride catalysts. A polar solvent such as N,N-dimethylformamide or acetonitrile was adopted as a reaction medium in order to dissolve the solid catalyst and mix it in TFEA and to lessen the temperature increase by diluting the heat from the exothermic reaction. Reaction conditions such as feed ratio, catalyst amount, solvent amount, reaction temperature, circulation flow rate of the gas phase reactant in the bubble reactor or the liquid phase reactant in the loop reactor have been optimized. The effect of reactor type on the product yield, byproducts, energy usage, and system controllability has been examined and compared.

144g Thermodynamics & Kinetics of the Erbium-Hydrogen System

Robert M. Ferrizz, Thomas J. Louck, M. Barry Ritchey, Loren I. Espada, Daniel R. Kammler, and Clark S. Snow. Sandia National Laboratories, PO Box 5800, Mail Stop 0878, Albuquerque, NM 87185

Metal hydrides are materials with a rich history and significant potential for emerging applications, including use as an energy carrier for hydrogen-based technologies[1,2]. Pressure-composition-temperature (PCT) properties and kinetics are critical attributes that help determine the suitability of a particular metal hydride system for a given application. In this report, the erbium-hydrogen system is examined experimentally, primarily using a PCT apparatus and thermal desorption spectroscopy, to characterize its isothermal P-C diagrams and hydriding/dehydriding kinetics. Rare earth metal hydrides, including erbium dihydride, are of recent interest due to unusual electronic and magnetic properties depending on hydrogen concentration[3]. A fundamental understanding of erbium-hydrogen isotherms and kinetics is important for accurately controlling composition following a range of processing steps, from loading films in a batch reactor to thermal vacuum treatment.

A well-calibrated PCT apparatus, using a volumetric technique known as Sieverts' method, can quantify the amount of hydrogen absorbed by the storage material. Simultaneously, a range of gauges are used to measure the equilibrium pressure. The resulting isotherms are useful for predicting the equilibrium composition for a given set of loading parameters. In this study, erbium-hydrogen isotherms have been generated for a range of operating temperatures. The equilibrium plateau pressures for the two-phase region are analyzed to determine differential enthalpy and entropy of reaction. For the α-β transition for a bulk erbium sample, ΔH and ΔS are determined to be -53.7 kcal/mol and -36.7 cal/molK. At present, the PCT has been used to verify and refine the accuracy of Lundin's isotherms for bulk erbium[4,5]. It is anticipated that experimental results for film samples may differ substantially from bulk samples, depending on film purity and processing parameters, particularly with respect to the lower β-phase boundary[3]. In addition to thermodynamic study, pressure versus time profiles generated using the PCT apparatus can be used to study isothermal hydriding kinetics. This analysis is ongoing at present.

Reaction kinetics control the composition of metal hydrides during non-equilibrium processes, such as the "cool down" stage of the loading process. Thermal desorption spectroscopy (TDS) is an experimental tool ideally suited for studying desorption kinetics. A TDS study[6] of erbium dihydride decomposition for film samples has determined that the activation energy for desorption is roughly 60.0 kcal/mol. In addition, this value has been found to vary considerably with a range of different processing parameters.

1. Latroche, M., Structural and thermodynamic properties of metallic hydrides used for energy storage. Journal of Physics and Chemistry of Solids 65, 517-522 (2004).

2. Sandrock, G., A panoramic overview of hydrogen storage alloys from a gas reaction point of view. Journal of Alloys and Compounds 293-295, 877-888 (1999).

3. Vajda, P., Hydrogen In Rare-Earth Metals, Including RH2+x Phases. Handbook on the Physics and Chemistry of Rare Earths 20, 207-291 (1995).

4. Lundin, C. E., The erbium-hydrogen system. Transactions of the Metallurgical Society of AIME 242, 903-907 (1968).

5. Lundin, C. E., Thermodynamics of the erbium-deuterium system. Transactions of the Metallurgical Society of AIME 242, 1161-1165 (1968).

6. Ferrizz, R. M., Erbium Hydride Thermal Desorption: Controlling Kinetics. Sandia National Laboratories report SAND2007-2659 (2007).

172f Kinetic Model of Complex Reaction Network: Methanol Steam Reforming

Tai-Shang Chen and Jia-Ming Chern. Department of Chemical Engineering, Tatung University, 40 Sec. 3, Chung-Shan N. Rd., Taipei, Taiwan

The reaction pathways and networks of many chemical reactions of industrial importance are usually so complicated that simple empirical rate laws are adopted to describe the reaction kinetics. Although empirical rate laws are able to correlate the kinetic data obtained from laboratory and/or pilot plant tests, the obtained rate laws and kinetic parameters cannot be used to design reactors with large scale-up ratio confidently. Therefore the proposed empirical kinetic models must be tested and revised in different scale-up stages with different reactor sizes. Such a stagewise scale-up process takes time, consumes material, and generates wastes. Direct scale-up from laboratory scale to full scale would be desired in highly competitive business environment. For safety, such a large-factor scale up requires a kinetic model derived from reaction mechanism that reflects all possible elementary steps. In this study, a methodology based on the Bodenstein approximation method and network reduction technique is used to derive the rate equations for methanol steam reforming. The experimental data published in the literature will be analyzed by various kinetic models of different reaction schemes. The obtained rate equations for methanol, hydrogen, carbon monoxide, carbon dioxide can be used to design reformer and some simulated design results and applications in fuel cell will be discussed.

205c Synthesis and Characterization of Au-Pt/SiO_2 Catalysts Prepared by Electroless Deposition Methods

Jayakiran Rebelli, Joseph H. Montoya, Melanie T. Schaal, Christopher T. Williams, and John R. Monnier. Chemical Engineering, University of South Carolina, 2C02, swearingen engineering center, 301 Main Street, Columbia, SC 29208

Electroless deposition (ED) has been used for the preparation of continuous, thin film metal coatings, but also has the potential for the preparation of novel, bimetallic catalysts by adjustment of the deposition parameters. Unlike traditional methods for the preparation of bimetallic catalysts, ED permits the selective deposition of a secondary metal (e.g., Au) on the surface of a catalytically active, primary metal surface (e.g., Pt). In order to prepare catalysts using this technique it is necessary to design a reactive, yet stable ED solution by the careful selection of the secondary metal salt and reducing agent (s), as well as the pH and temperature of the ED bath. In order to determine the stability and specificity of the bath, the ED solution is exposed to the support (e.g., SiO_2) to ensure that neither metal salt adsorption nor spontaneous reduction and precipitation of the secondary metal from solution occurs. For the case of Pt/SiO_2, a stable, room temperature solution was obtained at basic conditions (pH 8-10) for two different electroless baths. One bath used $AuCl_4^-$ as the metal source and HCHO as the reducing agent, while the second employed $Au(CN)_2^-$ and hydrazine, respectively. For both systems, predictive, sub-monolayer coverages of Au on Pt were achieved. Thus, a series of Au-Pt/SiO_2 catalysts was prepared and characterized using H2 chemisorption and transmission Fourier-transform infrared (FT-IR) spectroscopy of CO adsorption. Chemisorption results show a decrease in H_2 uptake on Pt as Au is added, indicating Au deposition on Pt has occurred. Changes in the population of adsorbed CO species with increasing Au content also corroborate this finding, as do XPS results. These catalysts have been initially evaluated for propylene hydrogenation to determine any obvious bimetallic ensemble effects. As expected, this hydrogenation reaction was determined to be structure insensitive as shown by the Pt-based, constant TOF values achieved for the different Au loadings. The efficacy of this ED approach for the synthesis of Au-Pt (and other Au-containing) bimetallic catalysts will also be discussed.

172b In-Situ, Fiber-Optic Measurement Techniques in Carbon Dioxide Expanded Liquid (CXL) Multiphase Reactors

Sean G. Mueller[1], Muthanna H. Al-Dahhan[2], and Milorad P. Dudukovic[2]. (1) Energy, Environmental, and Chemical Engineering, Washington University in St. Louis, 1 Brookings Drive, Campus Box 1180, Cupples II Hall, Room 207, St. Louis, MO 63130, (2) Energy, Environmental and Chemical Engineering, Washington University in St. Louis, 1 Brookings Drive, Campus Box 1180, Cupples II Hall, Room 207, St. Louis, MO 63130

Dense phase carbon dioxide, including liquid and supercritical CO2, has been gaining acceptance for potential use in industrial applications due to benefits of pressure-tunable density and transport properties, reduction of conventional solvent use, enhanced miscibility of reactants, optimized catalyst activity, and increased product selectivities, all of which decrease waste and pollution. CO2 expanded solvents (CXL's) also provide the benefit up of to 80% solvent replacement with a dense phase fluid such as carbon dioxide (Wei et al., 2002). However, analysis and modeling of expanded solvents and supercritical phase reactors are lacking. Also, physical properties of these mixtures are highly sensitive to changes in pressure, temperature, and composition. Therefore, a reliable understanding of phase behavior and critical phase behavior is necessary for both experimentation and modeling.

To gain a better understanding of phase behavior, as well as provide an online measurement tool, in-situ, fiber-optic probes have been developed to measure volumetric expansion (Mueller et al., 2007), to detect the phase transition from the subcritical to supercritical phase, and to characterize bubble dynamics in multiphase reactors. These properties are essential for modeling and determining phase separations, the amount of solvent and/or catalysts required as well as catalyst solubility. The objectives of this work are: 1) to advance the use of fiber optic probe measurement techniques into high pressure, high temperature, and multiphase reactors, 2) to quantify bubble dynamics in multiphase stirred tank reactors and 3) to quantify important phase behavior parameters in order to advance the fundamental understanding of CXL's.

References:

Wei, Ming; Musie, Ghezai T.; Busch, Daryle H.; Subramaniam, Bala. CO2-Expanded Solvents: Unique and Versatile Media for Performing Homogeneous Catalytic Oxidations. Journal of the American Chemical Society (2002), 124(11), 2513-2517.

Mueller, Sean G.; Werber, Jay R.; Al-Dahhan, Muthanna H.; Dudukovic, Milorad P. Using a Fiber-Optic Probe for the Measurement of Volumetric Expansion of Liquids. Industrial & Engineering Chemistry Research (2007), 46(12), 4330-4334.

69a Formation of Glycerol Derivatives by Catalytic Condensation and Acetalization

Aiguo Liu, Christopher Luckett, Hui Hui Li, and Wayne Turbeville. Sud-Chemie, Inc, 1600 W. Hill St., Louisville, KY 40232

Conversion of glycerol into high valued chemicals has been the focus of many recent researches due to exponential growth in biodiesel production in the world. Condensation to higher alcohols and acetalization with aldols to produce cyclic acetals are possible alternative routes to utilize low grade glycerol for fine chemicals. Catalysts and reaction conditions are critical factors that determine reaction pathways and products distribution. H-ZSM-5, Cu-ZSM-5 and CuCr based catalysts and different reaction media were selected for this research. Results indicate that higher alcohols (> C4) and various cyclic acetals, mainly the 1,3-dioxolane and its derivatives, were formed under various reaction conditions. ZSM-5 zeolite was found to promote the formation of cyclic acetals. Reductive atmosphere up to 68 bars of H2 did not significantly alternate the product distribution. The intra-molecular dehydration of glycerol is assumed to be an initial step leading to an aldol-condensation chain growth where methanol is likely involved. More selective catalysts will be needed to obtain certain targeted products.

Deep Oxidative Desulfurization of Fuels Catalyzed by Peroxophosphomolybdate Compounds in Ionic Liquids

Lining He[1], Huaming Li[1], Wenshuai Zhu[1], Junxiang Guo[1], Xue Jiang[1], Jidong Lu[2] and Yongsheng Yan[1]
[1] College of Chemistry and Chemical Engineering, Jiangsu University, Zhenjiang, China
[2] State Key Laboratory of Coal Combustion, Huazhong University of Science and Technology, Wuhan, China

Introduction

Recently, sulfur compounds in fuels have been a major source of air pollution and desulfurization of fuels is attracting wide attention. For environmental protection purpose, more stringent regulations and fuel specifications have been regulated[1]. It will be a trend to produce sulfur-free gasoline and diesel in next several years.

In the past several years, alternative deep desulfurization techniques have been extensively investigated, for example, oxidation[2], extraction[3], adsorption[4] and bio-processes[5]. Among these processes, oxidative desulfurization (ODS) has attracted much attention in recent few decades. ODS combined with extraction is considered as one of the most promising processes.

The room temperature ionic liquids (RTILs) are considered as "green solvents", environmental benign alternative solvents for catalysis, separations and so on. In recent papers, the desulfurization of sulfur compounds from fuels by extraction with RTILs has been reported[3]. However, the effect of sulfur removal is rather low due to the similarity between the sulfur-containing molecules and the remaining fuels.

The aim of this work is to develop a more efficient catalytic system (O/S=2) for desulfurization of sulfur-containing model oil. Quaternary ammonium salt with short carbon chain $[(C_4H_9)_4N]_3\{PO_4[MoO(O_2)_2]_4\}$ combining with H_2O_2 and ILs for catalytic oxidation-extraction desulfurization from model oil is described. With stoichiometric amounts of H_2O_2, the sulfur compounds could be almost completely removed from model oil.

Experimental

Preparation of peroxophosphomolybdates[6]

$[(C_4H_9)_4N]_3\{PO_4[MoO(O_2)_2]_4\}$ (**I**). To a solution of tetrabutylammonium bromide $[(C_4H_9)_4N]Br$ (1.0 g, 3.1 mmol) in 30% H_2O_2 (40 mL) was added phosphomolybdic acid hydrate $H_3PMo_{12}O_{40}\cdot 13H_2O$ (2.1 g, 1.0 mmol) in 30% H_2O_2 (10 mL), and the mixture was stirred vigorously at 40 ℃ for 4 h. The suspended mixture was cooled to room temperature until a white precipitate was produced, which was filtered off and washed with deionized water and ethanol and air-dried.

The preparation of the compounds $[C_{16}H_{33}NC_5H_5]_3\{PO_4[MoO(O_2)_2]_4\}$ and $[C_{14}H_{29}N(CH_3)_3]_3\text{-}\{PO_4[MoO(O_2)_2]_4\}$ was similar to the compound **I.**

Preparation of model oil

Dibenzothiophene (1.466 g, 7.793 mmol, 98%) was dissolved in a round-bottom flask (250 mL) with a solvent of *n*-octane (250 mL). The sulfur content of model oil containing DBT was 1000 ppm.

DBT oxidation

The ionic liquids were prepared by the published procedure[7, 8]. The oxidative and extractive desulfurization experiments of the model oil were carried out in a homemade 40 mL two-necked flask. 0.00156 mmol of catalyst [n(DBT)/n(catalyst) =100], 0.032 mL of 30 wt % H_2O_2 [n(H_2O_2)/ n(DBT) = 2] and the extracting solvent with IL (1 mL) was dissolved in the flask and then 5 mL of model oil (sulfur concentration was 1000 ppm) was added. The resulting mixture was stirred vigorously and heated to 70 ℃ in oil bath. The upper phase (*n*-octane) was withdraw and analyzed of sulfur content by GC-flame ionization detector (GC-FID). SE-54 capillary column (15 *m* x 0.32 *mm* inner diameter x 1.0 *um* film thickness).

Results and Discussion

The sulfur removal of different reaction systems are listed in Table 1. While [Bmim]BF_4 was used for desulfurization, the sulfur removal of model oil only reached 16.3%. Adding H_2O_2 as oxidant into the extraction system with [Bmim]BF_4 as extractant, the sulfur removal also only reached 31.5%. $[(C_4H_9)_4N]_3\{PO_4[MoO(O_2)_2]_4\}$ combined with H_2O_2 as desulfurization system, the effect of sulfur removal was also low, the sulfur removal values of 16.8% was given after chemical oxidation. However, in the case of $[(C_4H_9)_4N]_3\{PO_4[MoO(O_2)_2]_4\}$, H_2O_2 and [Bmim]BF_4 were employed together, the removal of DBT could reach 97.3%. From the result, we concluded that [Bmim]BF_4 was a benign solvent and it could obvious improve the activity of the short alkyl chain catalyst. This experiment clearly indicated a combination of short carbon chain catalyst and extractant with an oil-immiscible ionic liquid [Bmim]BF_4 could deeply remove DBT from model oil. The concentration of DBT in *n*-octane was reduced from 1000 to 91 ppm via catalytic oxidation with $[C_{14}H_{29}N(CH_3)_3]_3PO_4[MoO(O_2)_2]_4$ and H_2O_2, the sulfur removal reached 90.9%. Unexpectedly, after [Bmim]BF_4 was added to the system, the sulfur removal of DBT was decreased to 88.1%. Similar result was found in $[C_{16}H_{33}NC_5H_5]_3PO_4[MoO(O_2)_2]_4$, which obtained a sulfur removal of 81.3%. Steric hindrance of the long carbon chain became an obstacle for the approach of the sulfur atom to the catalytic active species in IL, leading to the lower S-removal.

Table 1. Sulfur removal of different desulfurization system in model oil[a]

Entry	catalyst $Q^+_3\{PO_4[MoO(O_2)_2]_4\}$	S-removal of different desulfurization system/%			
		catalyst+H_2O_2	**IL+catalyst+H_2O_2**	**IL**	**IL+H_2O_2**
1	$[(C_4H_9)_4N]^+$	16.8	97.3	16.3	31.5
2	$[C_{14}H_{29}N(CH_3)_3]^+$	90.9	88.1		
3	$[C_{16}H_{33}NC_5H_5]^+$	96.2	81.3		

[a] Model oil: 1000 ppm sulfur as DBT in n-octane; IL: [Bmim]BF_4; the different amphiphilic quaternary ammonium peroxophosphomolybdates as the catalysts ($Q^+_3\{PO_4[MoO(O_2)_2]_4\}$; Q = quaternary ammonium cations). Conditions: T = 70 ℃, t = 3 h, model oil = 5 mL, IL = 1 mL, [n(H_2O_2)/n(DBT)/n(catalyst) = 200:100:1].

Influence of different IL on DBT-removal

From the data of Table 1, We found that the desulfurization system of [Bmim]BF_4,catalyst and H_2O_2 could deeply remove DBT from model oil. To investigate the influence of different IL on desulfurization, five ILs [Bmim]BF_4, [Omim]BF_4, [Omim]TA, [Omim]PF_6 and [Bmim]PF_6 were used as extractants. The S-removal with different ILs in catalytic oxidation-extraction system was given in Table 2. The sulfur content of the model oil decreased from its original value of 1000 ppm to 27, 55, 93, 168 and 324 ppm in [Bmim]BF_4, [Omim]PF_6, [Bmim]PF_6, [Omim]BF_4 and [Omim]TA.

Because [Omim]BF_4, [Omim]PF_6 and [Bmim]PF_6 were immiscible with H_2O_2, the sulfur removal of model oil in these systems were lower than in the system using [Bmim]BF_4. In the system containing catalyst, H_2O_2 and [Omim]TA, the removal of sulfur probably was influenced by the anion of [Omim]TA. The data also indicate that a combination of catalytic oxidation and extraction with IL shows a high catalytic activity on removing DBT.

Table 2. Different ILs on DBT desulfurization[a]

Entry	**desulfurization system**	**sulfur removal (%)**
	IL + compound I + H_2O_2	
1	[Bmim]BF_4	97.3
2	[Omim]BF_4	83.2
3	[Bmim]PF_6	90.7
4	[Omim]PF_6	94.5
5	[Omim]TA	67.6

Reaction Condition : T = 70 ℃, t = 3 h, model oil = 5 mL, [n(H_2O_2)/n(DBT)/n(catalyst) =200 :100 :1], IL = 1 mL.

Conclusions

In this paper, we report that a combination of catalytic oxidation with a short carbon chain catalyst and extraction using [Bmim]BF_4 for the deep desulfurization of model oil under moderate conditions. This process is more superior to the mere solvent extraction with ionic liquid. The deep desulfurization system containing $[(C_4H_9)_4N]_3\{PO_4[MoO(O_2)_2]_4\}$, H_2O_2 and [Bmim]BF_4 can be recycled four times with an inconspicuous decrease in activity, which will be developed into a promising, "green" process of deep desulfurization.

Acknowledgments

This work was financially supported by the National Nature Science Foundation of China (No. 20676057, 20777029) and Jiangsu University Scientific Research Funding (No. 04JDG044).

References

1 Esser, J., Wasserscheid, P. and Jess, A. (2004), Deep Desulfurization of Oil Refinery Streams

by Extraction with Ionic Liquids, *Green Chem.*, June, pp. 316-322.

2 Lu, H. Y.; Gao, J. B.; Jiang, Z. X.; Jing, F.; Yang, Y. X.; Wang, G.; Li, C (2006), Ultra-deep Desulfurization of Diesel by Selective Oxidation with $[C_{18}H_{37}N(CH_3)_3]_4[H_2NaPW_{10}O_{36}]$ Catalyst Assembled in Emulsion Droplets, *J. Catal.*, March, pp. 369-375.

3 Bosmann, A.; Datsevich, L.; Jess, A.; Lauter, A.; Schmitz, C.; Wasserscheid, P. (2001), Deep Desulfurization of Diesel Fuel by Extraction with Ionic Liquids, *Chem. Commun*, November, pp. 2494-2495.

4 Yang, L. N.; Li, J.; Yuan, X. D.; Shen, J.; Qi, Y. T. (2007), One Step Non-hydrodesulfurization of Fuel Oil: Catalyzed Oxidation Adsorption Desulfurization Over HPWA-SBA-15, *J. Mol. Catal. A: Chemical.*, February, pp. 114-118.

5 Sipma, J.; Henstra, A. M.; Parshina, S. N.; Lens, P. N. L.; Lettinga, G.; Stams, A. J. M. (2006), Microbial CO Conversions with Applications in Synthesis Gas Purification and Biodesulfurization, *Crit. Rev. Biotechol.*, pp. 41-65.

6 Ishii,Y.; Yamawaki, K.; Ura, T.; Yamada, H.; Yoshida, T.; Ogawa, M. (1988), Hydrogen Peroxide Oxidation Catalyzed by Heteropoly Acids Combined with Cetylpyridinium Chloride. Epoxidation of Olefins and Allylic Alcohols, Ketonization of Alcohols and Diols, and Oxidative Cleavage of 1,2-diols and olefins, *J. Org. Chem.,* pp. 3587-3593.

7 Huddleston, J. G.; Visser, A. E.; Reichert, W. M.; Willauer, H. D.; Broker, G. A. and Rogers, R. D. (2001), Characterization and Comparison of Hydrophilic and Hydrophobic Room Temperature Ionic Liquids Incorporating the Imidazolium Cation, *Green Chem.*, July, pp. 156-164.

8 Bonhote, P.; Dias, A.P.; Papageorgiou, N.; Kalyanasundaram, K. and Gratzel, M. (1996), Hydrophobic, Highly Conductive Ambient-Temperature Molten Salts, *Inorg. Chem.*, pp. 1168-1178.

144a Dramatic Micellar Catalaysis on the Chracteristics of B-Z Oscillatory Reaction with Citric Acid

Umesh Kumar Utkoor, Chemistry, Osmania University, Hyderabad (A.P), India, Flat # 103,Ram's Residency, SBH C Colony, Saidabad, Hyderabad, 500 059, India and Chinna Rajanna Kamatala, Chemistry, Osmania University, Hyderabad (A.P) India, Plot #6, Road #2, Lalitha Nagar, Sri Sainagar Colony, Nagole(GSI), Hyderabad, 500 068.

A B-Z Oscillatory reaction with citric acid, Mn(II) and potassium bromate has been investigated in presence of sodium dodecyl sulphate(SDS),cetyl trimethyl ammonium bromide(CTAB) and Triton-X (Tx).Oscillatory parameters such as time of initiation (Ti), life time(Tl), Amplitude(A) and no of oscillations(n) have been dramatically decreased in presence of micellar media. The waves were reproduced with a perfect accuracy when conducted in teflon reactor. The concentration of controlling intermediate (Bromide ion) has been significantly affected by the adition of micelles. A plausible mechanism to account for the micellar effects on the oscillatory behavior has been proposed. Attempts have been made to simulate the oscillations using Runge-kutta method.

Oxidative coupling of methane on $BaTiO_3$ perovskites partially substituted with germanium

D.M.Boghaee[1],M.Hamidzadeh[2*], , S.Negin[2]

1- Department of Chemistry Sharif University of Technology P.O.Box: 11365-9516 -Tehran

dboghaei@sharif.edu1

2-National Petrochemical Company, Research and Technology Company, Iran Polymer & Petrochemical Research Institute, Pajohesh Blv., 15th km of Karaj Highway

hamidzadeh@nipc.net

Abstract

The perovskite oxide BaTiO3, doped with Ge oxides oxidized methane to C2 coupled hydrocarbons. The ultraviolet-visible spectra of pure barium titanate and Ge doped $BaTi_{(1-x)}\ Ge_xO_3$ (x=0.05 ,0.1) were compared. The pure barium titanate exhibited an ultraviolet absorption effect. The addition of GeO2 resulted in an increase in absorption of wavelength. Therefore, small amount of Ge decreased band gap energy. Oxygen is inserted by nucleophilic attack which may be considered as an acid-base process. For promoted $BaTiO_3$ catalysts with stable cation valance, electron donation ability of lattice oxygen is parallels to its ability of adsorb and active oxygen, and decrease the chance of deep oxidation of methyl radicals and C_2 hydrocarbons and thus will result in an increase of C_2 yield.

Keywords: methane, oxidative coupling, perovskite.

1. Introduction

The oxidative coupling of methane has received a great deal of attention for the last twenty five years Because the reaction has been understood as a promising route for the C2-hydrocarbon production from natural gas [1,2]. To identify suitable catalysts giving high C2-yield, a lot of catalysts have been examined for the oxidative coupling of methane. It has been reported that high surface basicity of a catalyst is necessary to enhance the C2 selectivity in the reaction [3]. Alkali metal oxides and alkaline earth metal oxides are strong bases so that they have catalytic activities for the oxidative coupling of methane. Alkali metal oxides seem to be unsuitable catalysts for the reaction because they form liquid phase of metal hydroxide in the presence of water vapor under reaction conditions [4], while alkaline earth metal oxides are suitable catalysts because they are easily dehydrated under reaction conditions[5].

In recent years much attention has been paid to nonmaterial composed of metals and semiconductors because they exhibit size-dependent optical, magnetic, electronic, and catalytic properties, and play an important role in the next generation of electronic devices [6]. Although the bulk properties of barium titanate (BT) ceramics have been widely investigated in the past years, more recently, there has been a renewal of interest in nano scale BT because its electrical properties are strongly dependent upon the grain size and the crystalline structure [7,8].

However, this method leads to the production of BT particles with un controlled and irregular morphologies, which limit the electrical properties of the resulting sintered ceramics. Other approaches such as homogeneous precipitation

[9–11] and sol–gel process [12–14] have been developed to improve the homogeneity and purity of BT powders. However, the products obtained from these solution-based techniques are either amorphous or represent precursor compounds, so that a calcination step at above 700 1C followed by milling is usually required to remove un reacted organics to form a pure crystalline phase. Thus a significant improvement in the powder quality is normally not achieved because the calcination and milling steps are similar to those used in the solid state reaction method. Furthermore, metal alkoxides used in sol–gel processes are generally too expensive for practical applications. Therefore, methods for cost-effective production of pure, ultra-fine nano crystalline BT particles with spherical morphology and narrow particle-size distribution, are to be developed.

. Various titanium precursors such as titanium oxide powder, titanium tetrachloride, titanium alkoxide, or a titanium oxide gel have been used. However, whether the particle size of the crystalline phase of the Ti precursors is the key factor remains unclear. The effects of the synthesis temperature, the pH value of the reaction medium, and the Ba/Ti ratio on the particle size and morphology have also been investigated extensively [16-18], but different results were reported by various researchers.

2.Experimental

Three kinds of $BaTi_{.95}Ge_{.05}O_3$ crystals prepared by the sol-gel technique using different solvent materials were examined in this study. One is BT crystals (sample A) synthesized by sol-gel technique using as-prepared titanium chloride, germanium oxide and barium hydroxide in propionic acid and mixed material refluxed for 3 hours. Sample B prepared with similar material but reflux time was 5 hours and solvent in sample C was 2- propanol. After cooling down to room and the reaction mixture was stirred for 30 minutes. The resulting mixture was heated without reflux to evaporate excess liquid until gel formed. The gel was dried at 130°C for 12 hours and calcined at 850°C for 6 hours.

Activity testing was carried out in a quartz fixed-bed micro-reactor (i.d. 9 mm) at ambient atmospheric pressure. 1.8 g catalyst (25-30 mesh) was loaded in the reactor and the length of the catalyst bed was approximately 30 mm. Quartz chips was filled in the remaining space of reactor so as to reduce the contribution from gas-phase reactions. The reactants, CH_4 (99.9%) and O_2 (99.9%) and N2(99.999%) as diluents, were co fed into the reactor. At the reactor outlet a cold trap was used to remove water from the exit gas stream. Blank runs with quartz chips showed negligible conversion at the reaction conditions. The reaction products were then analyzed with an on-line gas chromatograph equipped with a TCD, using a Poropak Q column for the separation of CH_4, CO_2, C_2H_4, and C_2H_6, and a 5A molecular sieve column for the separation of O_2, CH_4, and CO. The synthesis catalysts were tested in continuous flow reactor at 775°C. Then Sample D and E synthesis similar to sample A with new steqiometric

The crystallographic phase of the BT nano crystals was identified with XRD (PW3710 BASED) using CuKa radiation. A typical scan rate was 0.01° per second, and the 2θ range was 20–80°.

Absorption spectra were measured by a UV-vis spectrophotometer with a deuterium lamp. BaSO4 was used for calibration before test and as a reference sample during test.

The surface areas of calcined catalysts were determined by adsorption- desorption of nitrogen at liquid nitrogen temperature, using a Quantachrome ASAP 2010 accelerated surface area.

3. Results and discussion

The results of CH4 conversion and product selectivity of the $BaTi_{.95}Ge_{.05}O_{3\ catalysts}$ (Sample A,B, C) for OCM are listed in Table 1. Over the $BaTi_{.95}Ge_{.05}O_3$ catalysts, CH4 conversion and C2 selectivity are approximately 33- 40 and 33-48%, respectively. Propionic acid is better solvent in this reaction. When reflux time increased, probably defect in lattice decreased and band gap energy increased, so, conversion reduced.

Table1: The effect of solvent and reflux time on catalytic performance over $BaTi_{.95}Ge_{.05}O_3$ catalyst

	Reflux	Solvent	Conversion	Selectivity	Yield
Sample	Time		CH4	C2H4	C2
A	3	Propionic acid	38.5	48.3	18.6
B	5	Propionic acid	40	33.2	13.3
C	3	2-propanol	33.34	46.49	15.54

Ti^{4+} plays a dominating role in UV-Vis absorption in BaTiO3 derivative materials. The absorption edges of $BaTiO_3$, $BaTi_{.95}Ge_{.05}O_3$ and $BaTi_{0.9}Ge_{0.1}O_3$ are around 340, 390 and 400nm, respectively. So, Ge doping decreased band gap energy. A good correlation between C2 selectivity in OCM and the band gap energy of semiconductor can be seen (table2).This edge due to the optical transitions between 2p oxygen and 3d titanium states.. Increasing concentration of Ge increased band gap energy. In the diffractograms of E catalyst was observed GeO2 phase , so, a change in lattice structure of $BaGeO_3$ in sample A to GeO2 in sample E could be a result of high concentration of Ge in catalyst.

Table2: The effect of Ge on catalytic performance over $BaTiO_3$ catalysts

Sample	Materials	E_g(ev)	Yield%	Selectivity C2%	Conversion CH4	C2H4/C2H6
	BaTiO3	3.64	12.7	32.8	38.7	4.41
A	$BaTi_{.95}Ge_{.05}$	3.18	18.6	48.3	38.5	6.25
D	$BaTi_{0.9}Ge_{0.1}$	3.1	8.6	51.1	16.9	2.28
E	$BaTi_{0.85}Ge_{0.15}$	3.5	11.5	48.3	23.9	2.68

The smaller band gap of oxide, the more selective is the oxide in the OCM reaction. Because energy required to remove a valence electron and allow it to free conduct was decreased, so, conductivity was increased and neucleophilic oxygen was created easier.

At elevated reaction temperatures, these electron holes can leave bulk and become localized on the surface, the surface O^- ion (or localized electronic hole) can be responsible for hydrogen abstracting. Hydrogen abstracting from methane and regeneration of the reactive center requires removal of lattice oxygen from the lattice site which is expected to be highly endothermic. Blue shift of Ti-O vibration in FTIR spectrum in derivatives $BaTiO_3$ was seen (fig 1) with increasing of Ge concentration in catalyst. It would be the rate-limiting step in the catalytic cycle, so, conversion of methane would be decreased.

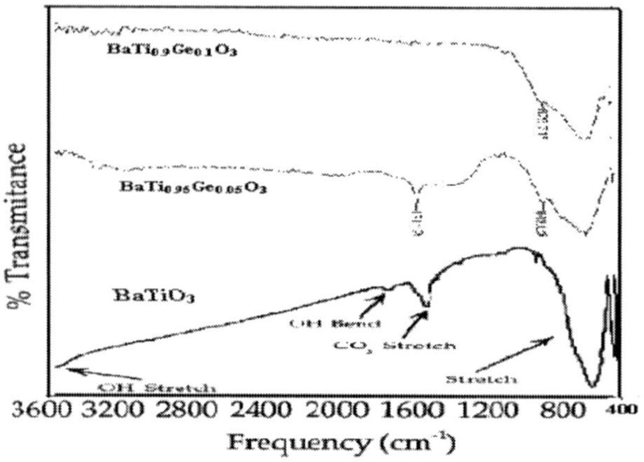

Fig1. FTIR spectra of $BaTiO_3$, $BaTi_{.95}Ge_{.05}O_3$ and $BaTi_{0.9}Ge_{0.1}O_3$

It is well known that diffusion of Ge cations on the M (Ti) site of the AMX3 perovskite lattice is slower than the A (Ba) site because the former produces larger lattice strain (such as 6-CN Ge^{4+}: 0.67A °, Ti^{4+}: 0.605A °) and hence needs higher energy incorporation than the latter. As a consequence, the separated BaGeO3 and BaTiO3 phases' are observed when sintering at 850°C and a solid solution of Ge in the BaTiO3 lattice requires higher sintering temperatures of Ge in the BaTiO3 lattice requires higher sintering temperatures. The apparent increase in a, c and V accompanying large XRD peaks shifts (Fig2) to the low angle of BT is attributed to the expansion of crystal cells resulting from the occupation of Ti sites with larger Ge ions in the BaTiO3 host lattice.

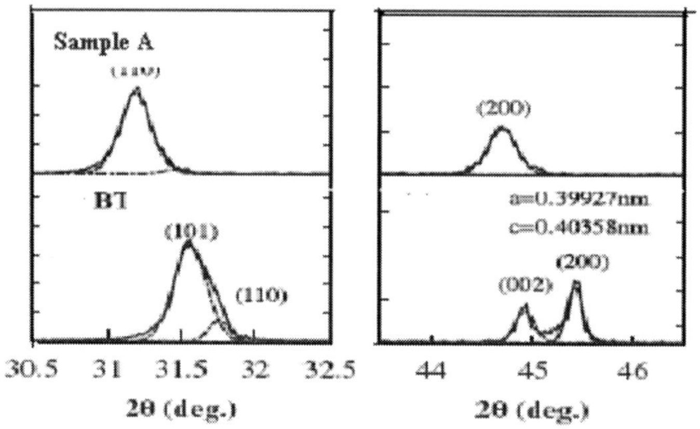

Fig2. Evolution of XRD peak BaTiO3, $BaTi_{.95}Ge_{.05}O_3$ The spectra were treated as: background removed, curve smoothed

The replacement of titanium by germanium revealed that for increasing Ge concentrations the conductivity in the n-type region decreases significantly. This can be attributed to the valence stability of Ge with respect to reduction.The packing of the ionic spheres illustrates the geometric relationship between the equilibrium (A – O) and (B – O) bond lengths; this relationship is referred to as the geometric tolerance factor *t*:

$$t = \frac{(A-O)}{\sqrt{2}(M-O)}$$

Where $t = 1$, the crystal structure is cubic and there is little elastic strain on the (A– O) or (M – O) bonds. Where $t < 1$, the (A – O) bonds are under tensile stress and (M –O) bonds are under compressive stress. The bonds' stress character is reversed if $t > 1$:the (A – O) bonds are under compressive stress whereas the (M – O) bonds are under tensile stress[19]. The (A – O) bond has a larger thermal expansion coefficient than the (M – O) bond, causing t to increase as the temperature is increased. Thus, the same series of structural phase-transitions that occur as a function of increasing tolerance factor also occur as a function of increasing temperature. In these catalysts t is larger than 1.

Where $t > 1$, no cooperative rotations of the oxygen-octahedra can relieve the tensile stress on the (M – O) bonds and compressive stress on the (A – O) bonds. To relieve the increased strain, the material crystallizes in a hexagonal-poly type phase at the expense of some of the electrostatic energy associated with the ionic structure. The replacement of titanium by germanium revealed that for increasing Ge concentrations the conductivity in the n-type region decreases significantly. This can be attributed to the valence stability of Ge with respect to reduction.

The surface areas were calculated according to the method of Brunauer, Emmett and Teller (BET), using 0.164 nm^2 as the cross-sectional area of the nitrogen molecule (table 3).

Table 3: Surface area of catalyst

Sample	BET surface area (m^2/g)	Total pore volume(ml/g)	Average pore radius(A)
A	72.9	0.092	25.4

4. Conclusion

Ge doped barium titanate, and decreased band gap energy, so, it increased oxygen interest to nucleophilic attack to methane and selectivity to C_2 products. Blue shift of Ti-O vibration decreased oxygen vacancy and methane conversion in reaction. Solvent is effect on band gap energy and performance of catalyst .

5. References

1. Keller, G. E.; Bhasin, M. M. *J. Catal.* **1982**, *73*, 9.

2. Lee, J. S.; Oyama, S. T. *Catal. Rev.-Sci. Eng.* **1988**, *30*, 249.

3. Lunsford, J. H. *Angew. Chem .* **1995**, *34*, 970.

4. Maiti, G. C.; Baerns, M. *Thermochimica Acta* **1995**, *261*, 69.

5. Byoung-Il An, Kwang-Hyun Ryu, Yong-Rok Kim, and Sung-Han Lee, *Bull. Korean Chem. Soc.* **2007**, 28, 6,1049.

6. S. Stupp, P.V. Braun, Science 277,**1997**, 1242.

7.Y. Yamamoto, K. Urabe, H. Banno, Jpn. J. Appl. Phys. 32 , **1993**, 4272.

8. S. Nourbaksh, I. Vasilyeva, J.N. Carter, Appl. Phys. Lett. 66 ,**1995**, 2804..

9. H.S. Potdar, P. Singh, S.B. Deshpande, P.D. Godbole, S.K. Date, Mater. Lett. 10 , **1990**, 112.

10. S. Kim, M. Lee, T. Noh, C. Lee, J. Mater. Sci. 31 , **1996**, 3643.

11. P. Gherardi, E. Matijevif, Colloid Surf. 32, **1988**, 257.

12. H.P. Beck, W. Eiser, R. Haberkorn, J. Eur. Ceram. Soc. 21 ,**2001**, 687.

13. H. Shimooka, M. Kuwabara, J. Am. Ceram. Soc. 79,**1996**, 2983.

14. G. Pfaff, J. Mater. Chem. 2, **1992**, 591

15.Y.S. Her, E. Matijevif, M.C. Chon, J. Mater. Res. 10, **1995**, 3106.

16.] M. Leoni, M. Viviani, P. Nanni, V. Buscaglia, J. Mater. Sci. Lett. 15 , **1996**, 1302.

17. P. Bendale, S. Venigalla, J.R. Ambrose, E.D. Verink Jr.,J.H. Adair, J. Am. Ceram. Soc. 76 , **1993**, 2619.

18. M. Viviani, J. Lemaitre, M.T. Buscaglia, P. Nanni, J. Eur. Ceram. Soc. 20 , **2000**, 315.

19. Jonathan Charles Denyszyn, Thesis for Doctor of Philosophy, THE DIELECTRIC BEHAVIOR OF PEROVSKITE-RELATEDMANGANESE OXIDES WITH STRETCHED BONDS OR MULTIFERROIC PROPERTIES, University of Texas at Austin, December **2006**

Ultra-Deep Desulfurization of Fuels Catalyzed by Molybdic Compound in Ionic Liquids

Wenshuai Zhu[1], Huaming Li[1], Jiexiang Xia[1], Xue Jiang[1], Jidong Lu[2], Yongsheng Yan[1] and Dan Xu[1]

[1] College of Chemistry and Chemical Engineering, Jiangsu University, Zhenjiang, China

[2] State Key Laboratory of Coal Combustion, Huazhong University of Science and Technology, Wuhan, China

Introduction

New processes for ultra-deep desulfurization are needed stringently as sulfur in fuels or oils has greatly contributed to the pollution of environment, such as, acid rain and air pollution[1]. With more and more stringent regulatory constraints, it is a trend to achieve little-to-no sulfur fuels (S-content< 10 ppm)[2]. It is difficult to achieve an ultra-low sulfur level because of some refractory sulfur compounds such as dibenzothiophene (DBT) and its derivatives through conventional hydrodesulfurization (HDS)[3]. Therefore, alternative desulfurization technologies are desired. Oxidative desulfurization(ODS) combined with extraction is considered one of the most promising among various processes.

Ionic liquids (ILs), as a novel green solvent, have been extensively employed in green chemistry instead of organic solvent because of low melting point, a wide liquid range, negligible vapor pressure and good solubility characteristics, etc. The new approach for desulfurization of diesel fuels by extraction with ILs has been recently described[4-11]. As a new desulfurization process, there is still much room for the development of more efficient method.

In this work, we employed commercially available molybdic compound as catalyst and developed extraction and catalytic oxidative desulfurization (ECODS) system composed of 1-butyl-3-methylimidazolium tetrafluoroborate ([bmim]BF_4), 30% H_2O_2 and $Na_2MoO_4.2H_2O$ for deep removal of DBT in model oil. The ionic liquids [bmim]BF_4 doped with the catalyst were selected for the ECODS. IL was immiscible with model oil, which not only served as extractant and reaction media but also made hydrogen peroxide stabilize in process of reaction. DBT was extracted from the model oil and oxidized in ionic liquid. These simple hydrophile molybdic compound not only possesses good catalytic activity but also hardly contaminates oil.

Experimental Section

Preparation of ionic liquid and model oil

The ionic liquids [bmim]BF_4, was synthesized as mentioned in the literature procedure[12]. Model oil was prepared by dissolving DBT, BT and 4, 6-DMDBT in n-octane, respectively. To give the corresponding sulfur content 1000, 1000 and 500 ppm.

Catalytic oxidative desulfurization of model oil

The oxidative desulfurization experiments were carried out in a home-made 40 mL two-necked flask. The mixture containing 5 mL of model oil, 0.064 mL of 30(wt.)% H_2O_2 [n(H_2O_2)/n(DBT)=4], 1 mL of IL and catalyst [n(S)/n(Mo)=20] was stirred vigorously at 70℃ for 3 h. After reaction was completed, the upper phase (model oil) was withdrawn and analyzed by gas chromatography (internal standard method nitrobenzene) coupled with flame ionization detector (GC-FID). A 15 *m* x 0.32 *mm* inner diameter x 1.0 *um* film thickness SE-54 capillary column was used for separation.

Results and Discussion

Effect of different desulfurization systems on removal of DBT in n-octane

Three different desulfurization systems of extraction, extraction coupled with chemical oxidation and extraction coupled with catalytic oxidation were conducted. The results were listed in Table 1. The catalyst and aqueous hydrogen peroxide could dissolve in IL and formed biphasic system, in which the oil phase was the upper and IL phase, along with catalyst and oxidizing agent, was the lower layer.

When $[bmim]BF_4$ was used as extractant for DBT-containing model oil at 30℃ and 70℃, the sulfur removal only reached 13.6% and 15.5% respectively. Addition of H_2O_2 in $[bmim]BF_4$, the sulfur removal of 31.5% was given after chemical oxidation. When $Na_2MoO_4.2H_2O$, H_2O_2 and IL were employed together, the removal of DBT increased sharply. This result indicated that IL played a significant role in the desulfurization system. In the case of the system comprising $Na_2MoO_4.2H_2O$, H_2O_2 and $[bmim]BF_4$, the sulfur content decreased from 1000 ppm to 10 ppm via extraction and catalytic oxidation, reaching 99.0% S-removal. However, the oxidative desulfurization system containing $Na_2MoO_4.2H_2O$ and H_2O_2 only led to 4.1% S-removal in the absence of IL, which contributed to a majority of hydrogen peroxide was decomposed under 70℃ in the presence of $Na_2MoO_4.2H_2O$. We concluded that IL not only served as extractant and reaction media but also made hydrogen peroxide stabilize at high temperature (70℃) with $Na_2MoO_4.2H_2O$ as catalyst. These results also indicate that the remarkable advantage of this process over the desulfurization of model oil by mere solvent extraction with IL or mere catalytic oxidation without IL.

Table 1 Comparison of different desulfurization systems of model diesel oil

Type of ILs	S-removal of different desulfurization system/%				
	IL		$IL+H_2O_2$	$IL+Na_2MoO_4+H_2O_2$	$Na_2MoO_4+H_2O_2$
	a	b			
$[bmim]BF_4$	13.6	15.5	31.5	99.0	4.1

Model oil was prepared by dissolving DBT in n-octane to give solutions with S-content 1000 ppm. Reaction conditions: T=70℃, t=3 h, model oil=5 mL, IL=1 mL, [n(DBT)/n(Catalyst)=20], $[n(H_2O_2)/n(DBT)]=4$ a: The mixture was stirred at 30℃ for 15 min. b: The mixture was stirred at 70℃ for 3 h.

Influence of the different of the substrates

To study the effect of the $Na_2MoO_4.2H_2O$ in $[Bmim]BF_4$ on the different sulfur compounds, the oxidation of three model sulfur compounds such as benzothiophene (BT), DBT and 4,6-dimethyldibenzothiophene (4,6-DMDBT) was carried out under the same conditions. As shown in Fig. 1, with the same reaction time, the sulfur removal through catalytic oxidation combined with extraction decreased in the order DBT>4, 6-DMDBT>BT. This order agreed well with the system of polyoxometalates/AcOH/H_2O_2[13]. The conversion of BT, DBT and 4, 6-DMDBT was 29.0%, 78.9% and 38.9% for 1 h, respectively. In the case of DBT, the sulfur removal could reach 99.0% in 3 h, however, the result unconspicuously continue improving after 3h. To BT and 4, 6-DMDBT, 49.6% and 69.5% S-removal were achieved in 3 h and the results continue increasing after 3 h. In this experiment, the sulfur conversion of BT and 4, 6-DMDBT could reach 61.0% and 89.5% in 8 h, respectively.

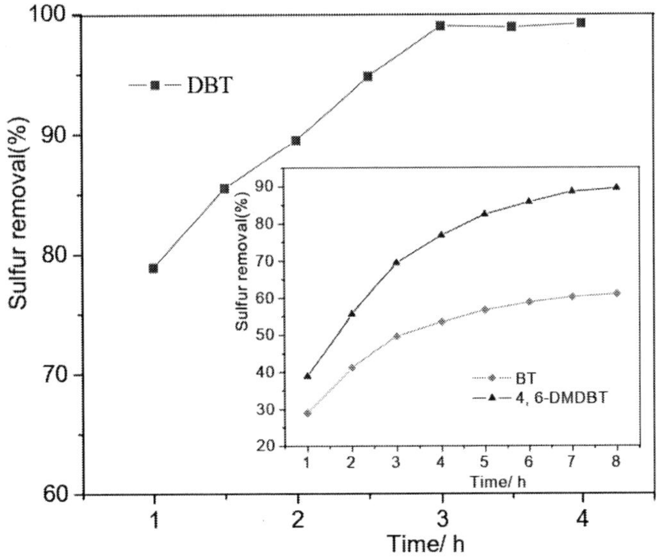

Fig. 1 Different substrates in the process of ECODS with $Na_2MoO_4.2H_2O$ as catalyst

Reaction conditions: T=70℃, model oil=5 mL, [Bmim]BF_4=1 mL, [n(S)/n(Catalyst)=20], [n(H_2O_2)/n(S) =4]

Conclusion

In this paper, commercially available molybdic compound dissolved in [bmim]BF_4 to oxidize BT, DBT and 4,6-DMDBT with hydrogen peroxide for deep desulfurization under moderate conditions. The sulfur removal of DBT-containing model oil can reach 99.0% under 70℃ for 3 h, which is the remarkable advantage of this process over the desulfurization by mere solvent extraction with IL or catalytic oxidation without IL. Moreover, catalyst hardly dissolved oil. It develops a simple, safe, reproducible and environmentally benign method for ultra-deep desulfurization.

Acknowledgments

This work was financially supported by the National Nature Science Foundation of China (No. 20676057, 20777029) and Jiangsu University Scientific Research Funding (No. 04JDG044).

References

1. Campos-Martin, J. M.; Capel-Sanchez, M. C.; Fierro, J. L. G. (2004), Highly Efficient Deep

Desulfurization of Fuels by Chemical Oxidation, *Green Chem*, October, pp. 557-562.

2. Esser, J.; Wasserscheid, P. and Jess, A. (2004), Deep Desulfurization of Oil Refinery Streams by Extraction with Ionic Liquids, *Green Chem.*, June, pp. 316-322.
3. Lu, H. Y.; Gao, J. B.; Jiang, Z. X.; Jing, F.; Yang, Y. X.; Wang, G.; Li, C (2006), Ultra-deep Desulfurization of Diesel by Selective Oxidation with $[C_{18}H_{37}N(CH_3)_3]_4[H_2NaPW_{10}O_{36}]$ Catalyst Assembled in Emulsion Droplets, *J. Catal.*, March, pp. 369-375.
4. Bosmann, A.; Datsevich, L.; Jess, A.; Lauter, A.; Schmitz, C.; Wasserscheid, P. (2001), Deep Desulfurization of Diesel Fuel by Extraction with Ionic Liquids, *Chem. Commun*, November, pp. 2494-2495.
5. Huang, C. P.; Chen, B. H.; Zhang, J.; Liu, Z.C.; Li, Y. X. (2004), Desulfurization of Gasoline by Extraction with New Ionic Liquids, *Energy Fuels*, September, pp. 1862-1864.
6. Nie, Y.; Li, C. X.; Sun, A. J.; Meng, H.; Wang, Z. H. (2006), Extractive Desulfurization of Gasoline Using Imidazolium-Based Phosphoric Ionic Liquids, *Energy Fuels*, August, pp. 2083-2087.
7. Zhang, S. G.; Zhang, Q. L.;.Zhang, Z. C. (2004), Extractive Desulfurization and Denitrogenation of Fuels Using Ionic Liquids, *Ind. Eng. Chem. Res.*, *December*, pp. 614-622.
8. Lo, W. H. H.; Yang, Y.; Wei, G. T. (2003), One-pot Desulfurization of Light Oils by Chemical Oxidation and Solvent Extraction with Room Temperature Ionic Liquids, *Green Chem*, *May*, pp. 639-642
9. Lu, L.; Cheng, H. F.; Gao, J. B.; Gao, G. H.; He, M. Y. (2007), Deep Oxidative Desulfurization of Fuels by Catalyzed Ioinic Liquid in the Presence of H_2O_2, *Energy Fuels*, 21, 383-384
10. Zhao, D. S.; Wang, J. L.; Zhou, E. P. (2007), Oxidative Desulfurization of Diesel Fuel Using a Bronsted Acid Room Temperature Ionic Liquid in the Presence of H_2O_2, *Green Chem.* 11, 1219-1222
11. Zhu, W. S.; Li, H. M.; Jiang, X.; Yan, Y. S.; Lu J. D.; Xia, J. X. (2007), Oxidative Desulfurization of Fuels Catalyzed by Peroxotungsten and Peroxomolybdenum Complexes in Ionic Liquids, *Energy Fuels*, 21, 2514-2516.
12. Huddleston, J. G.; Visser, A. E.; Reichert, W. M.; Willauer, H. D.; Broker, G. A. and Rogers, R. D. (2001), Characterization and Comparison of Hydrophilic and Hydrophobic Room Temperature Ionic Liquids Incorporating the Imidazolium Cation, *Green Chem.*, July, pp. 156-164.
13. Komintarachat, C.; Trakarnpruk, W. (2006), Oxidative Desulfurization Using Polyoxo-Metalates, *Ind. Eng. Chem. Res.*, February, pp. 1853-1856.

205d Comparison between Formic Acid Decomposition and CO Oxidation with Au Catalysts

Manuel Ojeda and **Enrique Iglesia**. Dept. of Chemical Engineering, University of California, Berkeley, CA 94720

Traditionally, gold has been envisioned a very poor active catalyst for formic acid decomposition because large Au particles display a high activation energy barrier for HCOOH adsorption. In this work, we have found that small Au particles (< 5 nm) are, however, much more active than Pt in the formation of CO2 and H2 from HCOOH. The reaction kinetics (zero-order in HCOOH partial pressure) and isotopic studies strongly suggest that formic acid decomposes on Au particles via a bimolecular reaction mechanism involving the participation of formate species and adsorbed HCOOH molecules.

The size of the gold particles did not change when the catalyst was treated under O2/He at different temperatures. The TEM images reveal that the Au particle size did not increase considerably as a result of the thermal treatment. However, the HCOOH decomposition rate decreased gradually as the treatment temperature was increased. Since gold particles sintering was not observed by TEM, our results suggest that very small (not detectable by TEM) Au species constitute the active sites. On the other hand, we have observed that the rate of CO oxidation in the presence of water is not affected significantly by the thermal treatment of the catalyst. In this case, our results suggest, therefore, that the active sites for the CO oxidation reaction correspond to the metallic Au particles (3-4 nm) observed by TEM.

Finally, we have also carried out a mechanistic study of the CO oxidation reaction with the Au catalyst. In particular, we have focused on the role of water presence and concentration on the reaction kinetics and mechanism. We have found that water presents is absolutely needed in order to attain a high activity and stability. Furthermore, the water effect depends on the partial pressure. These results highlight that a carefully control of water concentration is absolutely needed when testing the Au catalysts. In fact, this could be, at least partially, the reason of the many conflicting results that can be found in the literature about the water effect in the CO oxidation with Au-based catalysts.

Acknowledgements

L. Rendon, F. Ruiz and M. Avalos from "Centro de Ciencias de la Materia Condensada, UNAM, Mexico" are grateful acknowledged for the TEM images.

Development of Nanoscale Ceria-Based Sulfur-Tolerant Catalyst Formulations

Desikan Sundararajan and Abdul-Majeed Azad
Department of Chemical and Environmental Engineering
The University of Toledo, Toledo, OH 43606.

Abstract

The Diesel-based logistic fuel is the main artery of transportation by trucks and other locomotives in the continental US, as well as all across the globe. High temperature fuel cells running on jet fuel reformates are attractive to NASA and the US military due to their lightweight, zero emission and low acoustic signatures. Jet fuels, such as Jet A and JP-8 are already on board in many applications. These fuels, however, are invariably laden with sulfur, which if not removed, takes its toll on the performance and life-time of the reforming catalysts as well as on the SOFC anodes, by way of deactivation and poisoning. Thus, the development of sulfur-tolerant catalysts is critical to realize these goals. In this presentation, results of systematic structural and microstructural characterization on a series of ceria-based nanoscale sulfur-tolerant catalysts (with noble metal loading in the range of 0.33 to 1.0 wt. %) before and after sulfidation experiments conducted at 825°C for soak-time ranging from 4 to 24h, will be presented.

1. Introduction

In the light of impending energy crisis, ever increasing global demand for fuel and the quest for cleaner and greener energy, there is great interest in using logistic fuels (such as jet fuels, diesel and coal).The use of hydrogen-rich reformates of logistic fuels such as Avgas (aviation gasoline), JP-5, JP-8 and Jet-A are attractive as feeds for PEMFCs and/or SOFCs for NASA and the Army. NASA envisions employing fuel cells running on clean reformate from jet fuels in their futuristic un-manned aerial vehicles (UAVs), low emission alternate power (LEAP) missions, as well as trans-continental flights [1-3]. However, depending on the source and kind, jet fuels are invariably sulfur-laden. The sulfur content in these logistic fuels varies between 0.3 to 1 wt. %. Thus, fuel processors are required to convert jet fuels into hydrogen-rich reformate for extended periods in the presence of sulfur, and deliver hydrogen with little or no sulfur to the fuel cell stack. The presence of sulfur (mainly organosulfur) poisons and deactivates the reforming catalyst. Thus, to facilitate continuous uninterrupted operation of logistic fuel processors, robust sulfur-tolerant catalysts ought to be developed.

This work reports some preliminary results on the development, characterization and performance of nanoscale ceria-supported sulfur-tolerant catalysts with nominal loading of noble metals ($x_M \leq 1$ wt. %; M = mono-, bi- or tri-metal dispersion). These formulations were assessed in terms of their sulfur-tolerance and phase integrity under conditions of high sulfur exposure at temperatures typically employed in reforming.

2. Materials and Methods

Nanoscale gadolinia (Gd_2O_3)-doped ceria (CeO_2) - GDC - supports was synthesized via ammonia precipitation using appropriate water soluble precursors for cerium and gadolinium (from Alfa-Aesar, MA) followed by hydrothermal treatment of the basic slurry thus obtained, in an autoclave at 240°C, for 2h. The resulting material was dried overnight in a vacuum oven. Systematic structural and morphological characterizations were carried out to follow the crystallization and phase evolution as a function of temperature-time profile employed for calcination.

Catalysts were made by dispersing noble metals onto the GDC support material, using corresponding water soluble precursors: ruthenium (III) chloride (99.999%), palladium (II) nitrate (99.95 %), rhodium (III) nitrate hexahydrate (99.999 %), all from Alfa-Aesar. For this purpose, an aqueous slurry of GDC and the noble metal precursors was homogenized in a rotary evaporator for 4h, at the end of which, the thickened mass was dried overnight in a vacuum oven at 100°C. The dried cake consisting of soft agglomerates was crushed using mortar and pestle, and pulverized in a ball mill using 2-propanol and 5-mm diameter zirconia media. The slurry was dried under ambient conditions and sieved through 325 mesh screen. Each formulation was calcined in air at 700°C for 2h.

The sulfidation experiments were carried out by introducing 1000 ppm H_2S stream (ambient nitrogen) over 1g of catalyst at 825°C. To simulate the catalyst very close to the actual reforming conditions, the catalyst sample was first reduced in H_2 at 500°C for 1h. Sulfidation tests were performed on each catalyst for a period of 4h and 24h. Samples were sent to NSL Analytical, to evaluate the total amount of sulfur picked-up by each of these catalysts.

The catalysts were characterized by a host of techniques with respect to their crystallinity, active metal dispersion, structural and morphological features as well as their temperature-programmed reduction (TPR) characteristics. TPR provides a protocol to activate the catalyst prior to use. X-ray powder diffraction (XRD) signatures on the as-prepared and the post-sulfided sample were collected on a Philips diffractometer (PW 3050/60 X'pert Pro), using monochromatic $CuK\alpha_1$ radiations (λ = 1.54056Å) and Ni filter. The transmission electron microscopy was carried out using JEOL 3011 HR-TEM which is equipped with the capability of performing energy dispersive X-ray spectroscopy (EDS) as well. The total amount of sulfur retained on the catalyst after sulfidation experiments, was quantified by wet chemical techniques by NSL Analytical, Cleveland, OH.

3. Results and discussion

3.1 Rationale of using ceria-based supports

Typically, inert support materials are used to provide a large surface area to the catalyst. Alumina is a common support for a number of catalysts in oil refining due to its high thermal and chemical stability. Cerium oxide (ceria) is commercially used in automotive catalytic converters as a support for three-way catalyst for reducing harmful emissions from tail pipes of automobiles [4]. Ceria is well known for its high oxygen storage capacity (OSC) and easy reducibility due to its facile $Ce^{III} \leftrightarrow Ce^{IV}$ equilibrium [5-8]. CeO_2-based materials are quite active and highly stable catalysts both in reducing and oxidizing atmospheres in high-temperature regimes compared to other conventional support systems. This, in conjunction with a preparative technique that provides nanoscale particles, results in a high-surface-area catalyst with the possibility of high activity for the reforming of sulfur-laden fuels. When used as a support for a precious metal, it results in the transfer of oxygen to the supported metal while the support is then re-oxidized by the water present in the steam reforming reaction. Increased

OSC leads to increase in oxygen vacancies which contribute to increased reducibility of ceria as well. Although, CeO_2 is more active as a reforming catalyst than Ce_2O_3, this reduction process can provide a value addition property of sulfur tolerance to the catalyst.

Thermodynamic calculations show that in a reducing environment, Ce_2O_3 has higher sulfidation equilibrium constant than CeO_2, suggesting a possibility of formation of either cerium oxysulfide (partial sulfidation) or cerium sesquisulfide (complete sulfidation). This mechanism might help mitigate sulfur-led poisoning and deactivation in the long run. This sacrificial role may allow precious metal(s) to remain active longer, resulting in a longer on-stream life for the catalyst. As an additional benefit, it has been reported that H_2S adsorption on ceria is partially reversible. The thermodynamics of the CeO_2-H_2S reaction, however, do not allow the reduction of H_2S to below 200 ppm at about 627°C; even at a temperature of 827°C, the H_2S level cannot be reduced to below 100 ppm. This requires ceria to be in a reduced form, CeO_n ($n < 2$), to achieve the desired sulfur removal levels. After the reduction of the CeO_2 to Ce_2O_3, oxygen ion vacancies are created by the oxygen removal. As sulfur is introduced in the reduced ceria structure, sulfidation occurs rapidly. In the light of these characteristics, ceria was used as a support system for development of sulfur-tolerant catalysts [9-10].

3.2 Morphological and structural characterization

The TEM images of the as-synthesized catalysts R1B, and R1D are show in Fig. 1.

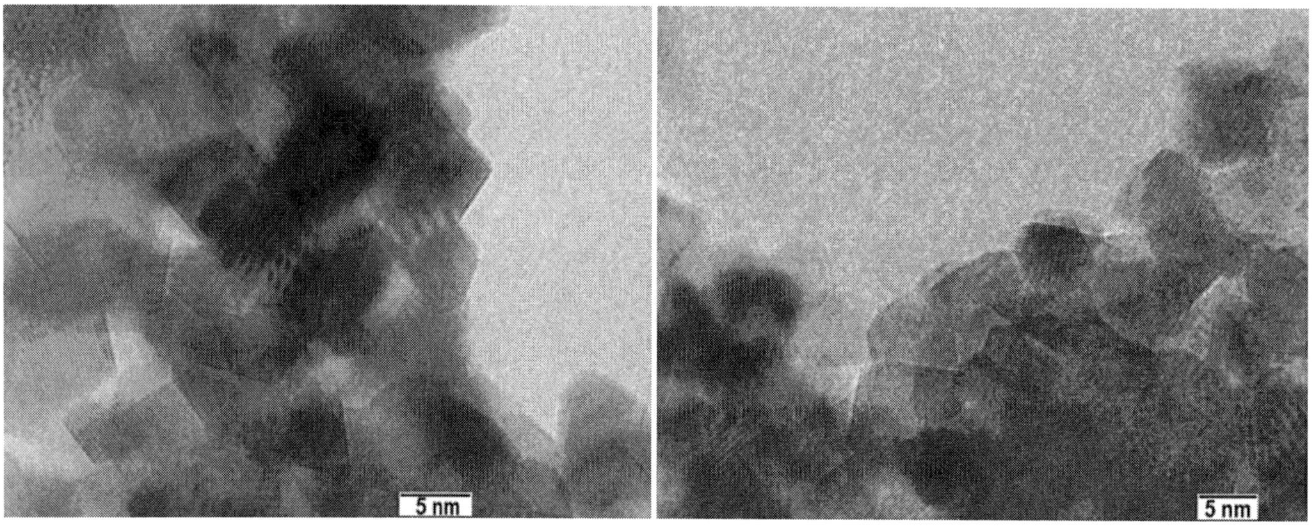

Fig. 1. TEM images of as-prepared catalysts R1B (left) and R1D (right).

As can be seen, both the catalysts possess nanofeatures. The contrasting light and dark particle contours belong to the GDC support and the noble metal, respectively. They further suggest that the dispersion of the noble metals (in very low concentrations) is highly uniform throughout the matrix. Thus, it is evident that the preparatory technique utilized for making these materials is quite effective for yielding uniform distribution of the noble metals across the catalyst support.

The nanofeatures observed in the TEM images are also corroborated by the rather broad diffraction peaks in the XRD patterns shown in the Fig. 2.

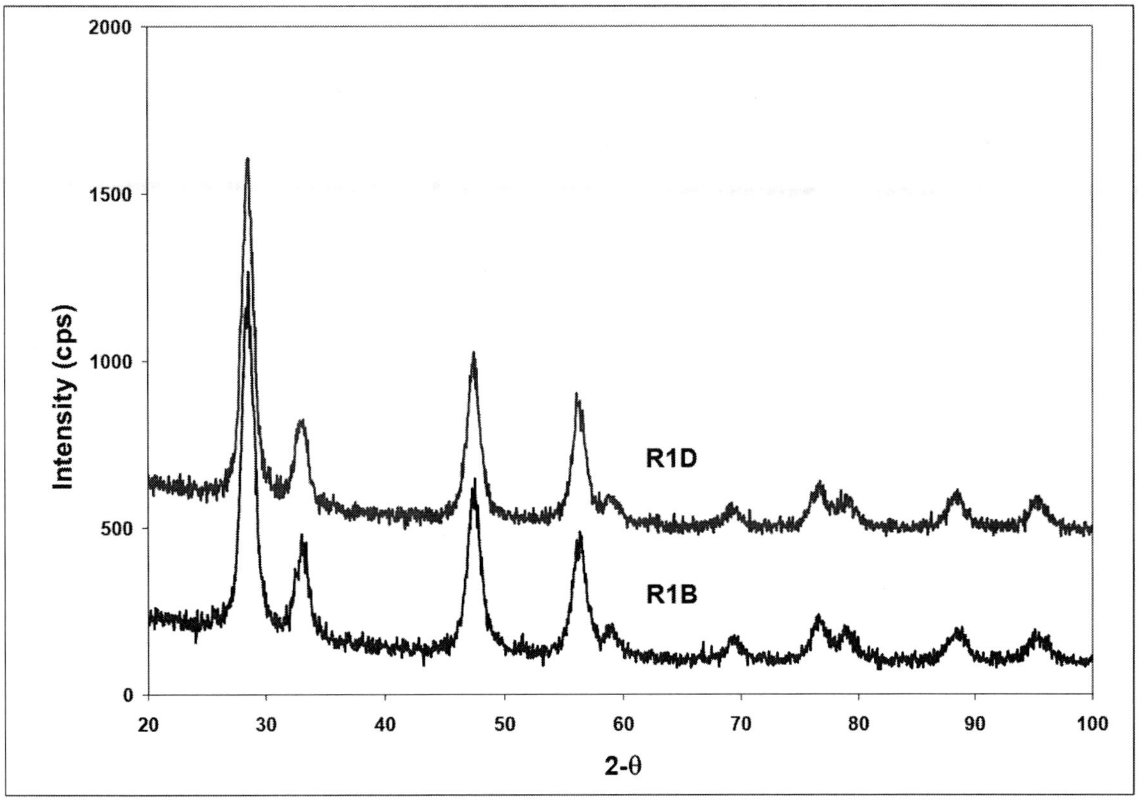

Fig. 2. XRD signatures of the calcined catalyst powders.

Estimation of the crystallite size in the powders using Scherrer's equation, yielded values between 7 and 8 nm, which is in excellent agreement with those seen in the TEM images. The two XRD patterns also correspond to the standard ICDD pattern for the 10 mole% gadolina-doped ceria (ICDD card # 75-0162). Peaks for noble meals are not discernable due to their low loading ($\leq$ 1 wt. %), which is below the detection limit of the XRD technique. The nanofeatures of these formulations ensure large surface area and effective noble metal dispersion. This lowers the loading and keeps the manufacturing cost low without compromising the performance.

3.3 Evaluation of sulfur tolerance

As discussed earlier, sulfur tolerance of a catalyst is a prime requirement of logistic fuel reforming. Recent studies have shown irreversible sulfur poisoning as the main reason for catalyst deactivation [11]. Thus, it is imperative that the catalysts intended for the reforming of sulfur-laden fuels be evaluated for their sulfur tolerance. For this purpose, the formulations synthesized in this work were tested for their sulfur capacity and phase integrity under conditions of sulfur exposure for 4-24h at high temperatures.

The cumulative sulfur sorbed by the catalysts after exposure to a gas stream containing 1000 ppm in nitrogen for 4 and 24h is shown in Fig. 3. In order to show the effect of variation of the noble metal loading, the sulfur sorption characteristic of R1A (single noble metal loading) is also included.

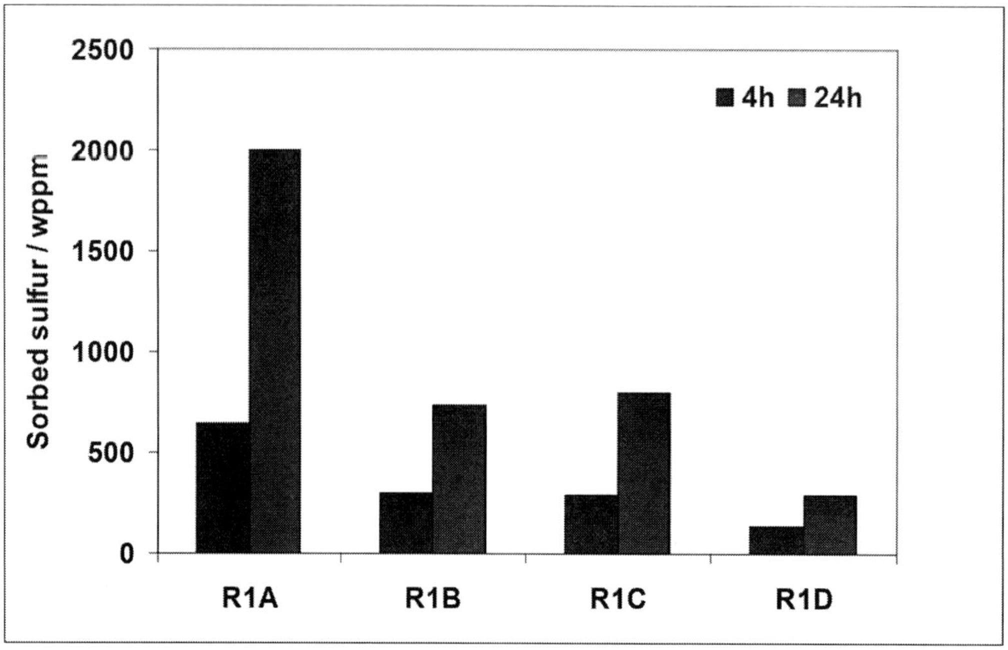

Fig. 3. Sulfur-sorption characteristics of various catalysts synthesized in this work.

As can be seen, the amount of sulfur sorbed was considerably lower in the case of bi-metallic (R1B and R1C) formulations than in the case of R1A (single metal dispersion), and further decreased in the case of tri-metallic formulation (R1D). Figure 4 shows the TEM images of the post-sulfided R1B sample.

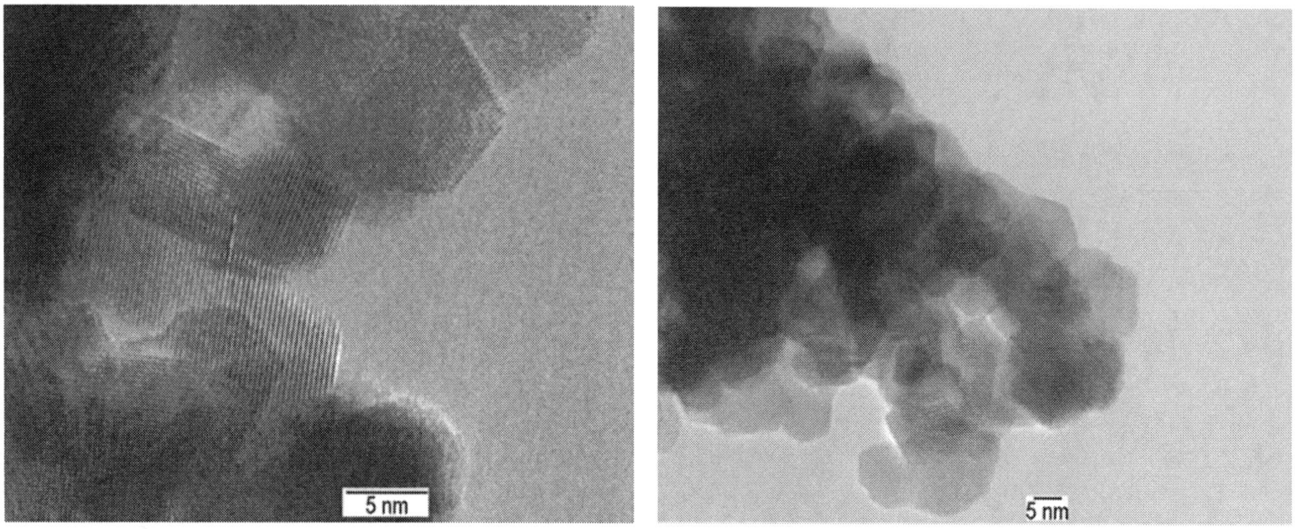

Fig. 4. TEM images of catalyst R1B sulfided for 4 (left) and 24h (right) at 825°C in 1000ppm H_2S/bal. N_2 stream.

As can be seen, there is evidence of slight grain growth in the sample sulfided for longer (24h) duration; the particles have grown to about 10-12 nm in size. This is corroborated by the concomitant narrowing of the diffraction peaks; the XRD patterns of R1B sample before and after sulfidation are compared in Fig. 5.

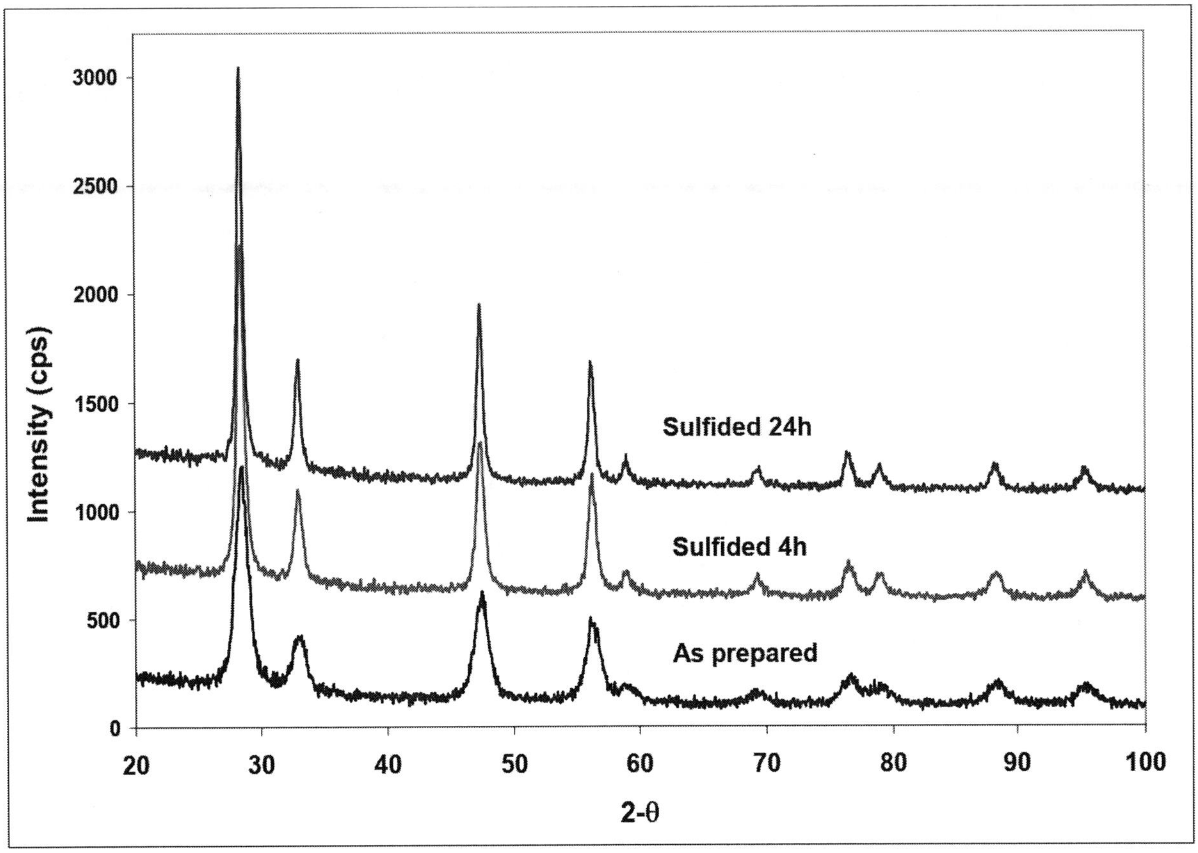

Fig. 5. Comparative XRD patterns of R1B.

The peak sharpening is an indication of systematic crystallinity enhancement as a function of time at high temperature (825°C); peak narrowing is an artifact of grain growth and crystallite size enhancement upon sulfidation. However, the Scherrer equation calculations and the TEM evidence both suggest that the grain growth in the sulfided samples is only nominal, indicating no significant surface area reduction. It can also be seen that no new phases could be discerned, suggesting phase integrity of the catalysts under sulfidation at high temperatures.

4. Conclusions

Nanoscale doped-ceria support were synthesized and used in the development of a number of novel sulfur-tolerant catalyst formulations. The morphological and sulfidation characteristics of the catalyst loaded with one, two and three noble metals were examined. It was found that all the formulations exhibited sulfur-tolerance at 825°C for on-stream time varying between 4 and 24h, without significant reduction in active surface area of the formation of additional phases. A systematic decrease in sulfur pick-up was believed to be due to the mechanism of "cooperative synergy" among the noble metals. Steam-reforming experiments with sulfur-laden simulated jet fuel streams are in progress.

Acknowledgement

Financial support by the US Army (contract # W909MY-06-C-0048) is gratefully acknowledged.

References

1. A. Liang, Emerging fuel cell developments at NASA for aircraft applications, presented at the 4th Annual SECA Meeting, Seattle, WA, USA, April, 15–16, 2003.
2. C. Song, *Catalysis Today*, **77** (2002) 17-49.
3. M. Krumpelt, T.R. Krause, J.D. Carterm, J.P. Kopasz and S. Ahmed, *Catalysis Today*, **77** (2002) 3-16.
4. A. Trovarelli (Ed.), *Catalysis by Ceria and Related Materials*, vol. 2, Imperial College Press, London, 2002.
5. M. Mogensen, N.M. Sammes and G.A. Thompsett, *Solid State Ionics*, **129** (2000) 63-94.
6. H.C. Yao, *J. Catal.* **86** (1984) 254-265.
7. M. Funabiki, T. Yamada and K. Kayano, *Catalysis,* **10** (1991) 33-43.
8. S. Tagliaferri, R.A. Köppel and A. Baiker, *Ind. Engg. Chem. Res.* **38** (1991) 108-117.
9. A.-M. Azad and M.J. Duran, *Applied Catalysis A: General,* **330** (2007) 77-88.
10. A.-M. Azad, M.J. Duran, A.K. McCoy and M.A. Abraham, *Applied Catalysis A: General,* **332** (2007) 225-236.
11. Y. Lu, J. Chen, Y. Liu, Q. Xue and M. He, *J. Catalysis,* **254** (2008) 39-48.

172c Flow Distribution in a High Pressure Trickle Bed Reactor

Zeljko V. Kuzeljevic, MIlorad P. Dudukovic, and Muthanna H. Al-Dahhan. Energy, Environmental and Chemical Engineering, Washington University in Saint Louis, One Brookings Drive, Saint Louis, MO 63130

Trickle Bed Reactors (TBRs) are the multiphase reactors in which gas and liquid phases flow co-currently down a fixed bed of catalyst. These reactors are used in the petroleum and petrochemical processes, such as hydrodesulphurization and hydrogenation, but also find application in the oxidation of organics in the wastewater effluents, volatile organic compound abatement in the air pollution control, and enzymatic reactions (Al-Dahhan et al., 1997; Sie and Krishna, 1998). Most commonly, these reactors operate under elevated pressure that can go up to 300 bars.

The performance of TBRs is crucially dependent on the liquid phase distribution across the packing. Poor distribution (liquid mal-distribution) will decrease the conversion and can lead to hot spots.

In this study, we experimentally examined the hydrodynamics of a high pressure TBR. Hydrodynamics and flow distribution in an air-water, 6" column system is quantified in terms of overall (liquid holdup and pressure drop) and distributed (effluent fluxes of both gas and liquid phase) parameters.

References:

Al-Dahhan, M. H., F. Larachi, M. P. Dudukovic and A. Laurent (1997). "High-Pressure Trickle-Bed Reactors: A Review." Industrial & Engineering Chemistry Research 36(8): 3292-3314.

Sie, S. T. and R. Krishna (1998). "Process development and scale up: III. Scale-up and scale-down of trickle bed processes." Reviews in Chemical Engineering 14(3): 203-252.

144c Activated Carbon Supported Catalysts and Adsorbents for WGS Reaction

Burcu Selen-Çaglayan and **A.Erhan Aksoylu**. Chemical Engineering, Bogaziçi University, 34342 Bebek, Istanbul, Turkey

Activated carbon supported catalysts were prepared for Water Gas Shift (WGS) reaction. The catalysts are all bimetallic and have Pt and Au as the primary active metal for high temperature and low temperature WGS reactions (HTS and LTS), respectively. The reaction tests were conducted for tempertaure range of 200-275 oC for LTS and 300-400 oC for HTS. The results indicated that (i) the AC pretreatment procedure has a direct influence on the properties and the activity/stability characteristics of the catalysts, and (ii) bimetallic catalysts, especially Pt-Ni for HTS and Au-Ce for LTS, has superior activities compared to the monometallic reference catalysts. In the second part of the study, several AC based adsorbents were designed through changing the surface chemistry of AC by various oxidation methods. The CO2 adsorption capacity and adsorption capacity-AC surface chemistry relation was investigated. In the final stage of the study, those adsorbents were used as a physical mixture with the catalysts in the reactor aiming to shift the WGS equilibrium for higher H2 production.

Design Implications of the Nonlinear Dynamics of the Monolithic Loop Reactor for Fischer-Tropsch Synthesis

Pietro Altimari[a], Costin S. Bildea[b], Ruud van Ommen[a], Johan Grievink[a]

[a]*Delft University of Technology, Julianalaan136, 2628BL Delft, The Netherlands*

bUniversity Politehnica Bucharest, Polizu 1, RO-011061Bucharest, Romania

Abstract

The nonlinear behavior of the monolith reactor with a liquid phase recycle for Fischer-Tropsch synthesis is investigated. A dimensionless distributed heterogeneous model is developed to describe reactor dynamics and bifurcation analysis is performed by varying the coolant temperature, the reactor gas inlet flow rate and the effectiveness factor of the external heat exchanger. As system parameters vary, coexisting stationary patterns are observed. Stability analysis of the reactor steady state solution regimes is carried out and Hopf bifurcation points are identified. As a result, periodic spatiotemporal patterns are found to arise in wide regions of the parameter space. The evolution of periodic solution regimes is examined by means of numerical simulation of the reactor model. Implications for operation and control are discussed.

Keywords: Fischer-Tropsch, monolithic reactor, state multiplicity, Hopf bifurcations.

Introduction

The emerging evidence of actually declining supply of crude oils worldwide and the increasing demand of petroleum products have lead, during the last few years, to a dramatic growth of crude prices. Motivated by this, considerable efforts are nowadays provided in detecting reliable solutions to convert coal, natural gas and biomass to liquid fuels. In this scenario, renewed interest is devoted to Fischer-Tropsch synthesis. This is a process to convert synthesis gas (1:2 mixture of CO and H_2) to liquid fuels and base chemicals. Synthesis gas can be obtained, for example, by steam reforming of methane, making Fischer Tropsch synthesis a viable solution to exploit available natural gas reserves for liquid fuels generation.

While being an old process, reactor design for Fischer Tropsch synthesis has been always a non optimal compromise. Traditional reactor solutions as fluidized beds, slurry bubble reactors and fixed bed reactors have disadvantages such as a high selectivity to lower alkanes, large backmixing, and poor heat removal, respectively. In this framework, a monolith loop reactor has been recently shown to provide a promising alternative for Fischer-Tropsch synthesis [1]. According to the proposed solution, a large fraction of the reactor outlet liquid flow rate is cooled by an external heat exchanger and recycled to the reactor in order to achieve reactor temperature control (Fig.1). Performances of a reactor of competitive size have been examined showing that high productivity and selectivity towards heavy hydrocarbons can be achieved [2]. Nevertheless, controllability issues need to be thoroughly investigated in order to assess the feasibility of the proposed reactor solution.

In this contribution, the nonlinear behavior of the monolith loop reactor for Fischer-Tropsch synthesis is investigated. A dimensionless distributed heterogeneous model is formulated to describe reactor dynamics and bifurcation analysis is performed as the coolant temperature, the reactor gas inlet flow rate and the effectiveness factor of the external heat exchanger are varied.

The paper is structured as follows. In section 2, the mathematical model used to describe the dynamic behavior of the monolith loop reactor is presented in dimensionless form. Results of the bifurcation analysis of the described model are reported in section 3. Final remarks end the paper.

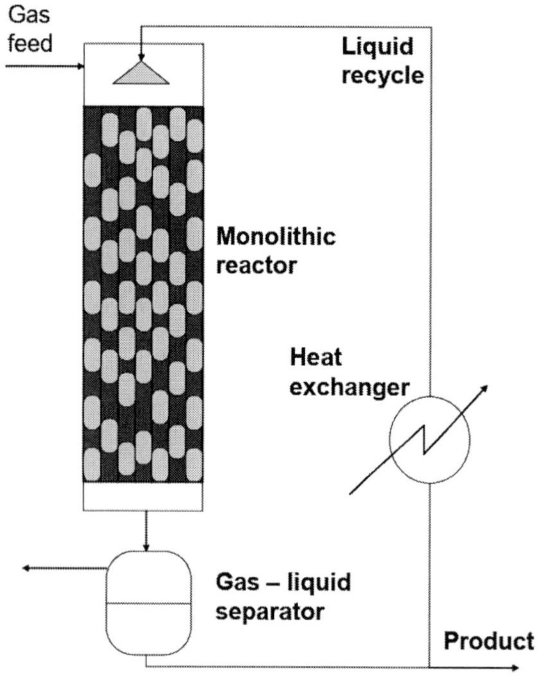

Figure 1.Flowsheet of thee monolith loop reactor

Mathematical model

Monoliths are ceramic blocks of parallel and straight channels. This open structure hardly obstructs the flow, guaranteeing low pressure drops. Catalyst can be deposited on the walls of the channels in the form of a thin layer leading to short diffusion distances. Extensively used, over the past decades, as automotive exhaust converters, monoliths have been recently proved to offer a promising alternative for gas-liquid-solid applications [3]. In these systems, a two-phase flow pattern, called Taylor flow regime, arises at sufficient superficial velocity. This hydrodynamic regime is characterized by the consecutive flow of bubbles of gas and liquid slugs through the channels (Fig. 1). Under these conditions, mass transfer is favored as result of the fast circulation within the liquid slug.

In this section, a mathematical model is presented describing the dynamic behavior of the monolith loop reactor for Fischer-Tropsch synthesis. Mass balances are formulated for the reactants, carbon monoxide and hydrogen, and products, water and hydrocarbons, in the gas and liquid phase (Eq. (1), (2)). The reaction of hydrogen and carbon monoxide to water and hydrocarbon causes a strong contraction of the gas-phase, leading to a significant decrease of the gas velocity. For this reason, the gas velocity is modeled as variable and an overall mass balance for the gas phase is defined to describe the evolution of the gas volume fraction along the reactor length (Eq. (3)). Since the amount of liquid hydrocarbons produced by the reaction is small compared to the recycled liquid stream, the liquid velocity is assumed to be constant along the reactor. Mass balances for both reactants and products in the solid phase (Eq. (4)) are formulated assuming that concentration gradients within the catalyst layer are negligible.

An overall energy balance is defined to describe the evolution of the reactor temperature along the reactor (Eq. (5)). Since the heat capacity of the gas is negligible, only the liquid and the solid phase are considered in the formulation of the energy balance. Pressure along the reactor is assumed to be constant. The synthesis gas consumption rate (Eq. (6)) is described using the expression given by Yates and Satterfield [4].

Dirichlet boundary conditions are used for all the state variables (Eq. (7)-(9)). In particular, the reactor inlet temperature is expressed as convex combination of the coolant and reactor outlet temperatures (Eq. (9)).

Following the previous assumptions, dimensionless mass and energy balance equations for the monolith loop reactor write as follows:

Species mass balances in the gas phase:

$$C_{g,i}\frac{\partial \varepsilon}{\partial t^*}+\varepsilon\frac{\partial C_{g,i}}{\partial t^*}=-C_{g,i}\frac{\partial U_g}{\partial \xi}-U_g\frac{\partial C_{g,i}}{\partial \xi}-(w_{g,l})_i\left(\frac{C_{g,i}}{m_i}-C_{l,i}\right)-(w_{g,s})_i\left(\frac{C_{g,i}}{m_i}-C_{s,i}\right) \quad (1)$$

Species mass balances in the liquid phase:

$$(1-\varepsilon)\frac{\partial C_{l,i}}{\partial t^*}-C_{l,i}\frac{\partial \varepsilon}{\partial t^*}=-U_l\frac{\partial c_{l,i}}{\partial \xi}+(w_{gl})_i\left(\frac{C_{g,i}}{m_i}-C_{l,i}\right)-(w_{ls})_i\left(C_{l,i}-C_{s,i}\right) \quad (2)$$

Overall mass balance in the gas phase:

$$-\frac{\varepsilon}{(\theta+1)}\frac{\partial \theta}{\partial t^*}+\frac{\partial \varepsilon}{\partial t^*}=\frac{U_g}{(\theta+1)}\frac{\partial u_g}{\partial \xi}-\frac{\partial U_g}{\partial \xi}-(\theta+1)\sum_i\left[(w_{g,l})_i\left(\frac{C_{g,i}}{m_i}-C_{l,i}\right)+(w_{g,s})_i\left(\frac{C_{g,i}}{m_i}-C_{s,i}\right)\right] \quad (3)$$

Species mass balances in the solid phase:

$$\frac{\partial C_{s,i}}{\partial t^*}=\frac{1}{\lambda}\left[\left(w_{g,s}\right)_i\left(\frac{c_{g,i}}{m_i}-c_{s,i}\right)+(w_{l,s})_i\left(c_{l,i}-c_{s,i}\right)\right]-v_i\,Da\,R(C_{s,H_2},C_{s,CO},\theta) \quad (4)$$

Overall energy balance in the gas-phase:

$$(Le-\varepsilon)\frac{\partial \theta}{\partial t^*}-(\theta+1)\frac{\partial \varepsilon}{\partial t^*}=-U_l\frac{\partial \theta}{\partial \xi}+\cdot\lambda\cdot Da\cdot B\cdot R(C_{s,H_2},C_{s,CO},\theta) \quad (5)$$

Reaction rate and boundary conditions:

$$R(C_{s,H_2}, C_{s,CO}, \theta) = \frac{(\theta+1)^2 \cdot \exp\left(\frac{\gamma_a \theta}{\theta+1}\right) C_{s,H_2} \cdot C_{s,CO}}{\left(1 + b_r \cdot (\theta+1) \cdot \exp\left(-\frac{\gamma_b \theta}{\theta+1}\right) C_{s,CO}\right)^2} \quad (6)$$

$$C_{g,i}(0,t) = C_{g,i,0};\ C_{l,i}(0,t) = C_{l,i,0};\ U_g(0,t) = U_{g,0} \quad (7)$$

$$\theta(0,t) = \theta_0 \mu_{HE} + \theta(1,t)(1 - \mu_{HE}) \quad (9)$$

The dimensionless variables appearing in the model are the axial coordinate $\xi = x/L$, the time $t^* = u_0 t/L$, the concentration $C_{i,j} = c_{i,j}/c_0$ of the specie j in the phase i, the temperature $\theta = (T - T_0)/T_0$, the gas volume fraction $\varepsilon = A_g/A$ in the channel, the mass transfer coefficient $(w_{i,j})_k = (k_{i,j}a_{i,j})_k\, L/u_0$ for the transport of the specie k between the phases i and j, the Damkholer number $Da = L\rho_s a_0 exp(-T_a/T_0) C_0 (RT_0)^2/u_0$, the adiabatic temperature rise $B = \Delta H C_0/\rho_l c p_l T_0$, the Lewis number $Le = (\rho_l c p_l A + \rho_l c p_l A_s)/\rho_l c p_l A$, the superficial gas velocity $U_g = u_g/u_0$, the superficial liquid velocity $U_l = u_l/u_0$, the heat exchanger effectiveness factor $\mu_{HE} = 2US/(US + 2F_l \rho_l c p_l)$, the solubility coefficient $m_i = HM_l/\rho_l RT_0$ of the specie i in the liquid phase, the activation energies $\gamma_a = T_a/T_0$ and $\gamma_b = T_b/T_0$, the ratio $\lambda = A_s/A$ between the solid and the gas-liquid sections, the kinetic constant $b_r = b_0 exp(T_b/T_0)$. Physical properties of the liquid and gases solubility can be found in [5, 6]. Under Taylor flow conditions, the gas volume fraction is related to the superficial velocities by $\varepsilon = U_g/(U_g + U_l)$. Replacing this into Eq. (1)-(9) gives rise to a set of partial differential equations in the unknowns $C_{g,i}(\xi,t)$, $C_{l,j}(\xi,t)$, $C_{s,j}(\xi,t)$, $U_g(\xi,t)$, $\theta(\xi,t)$.

Nonlinear analysis

In this section, the nonlinear behavior of monolith loop reactor is described as the reactor inlet gas velocity, the coolant temperature, and the effectiveness factor of the external heat exchanger are varied. Results concerning the steady state behavior of the system are reported in subsection 3.1, while the stability of the stationary solutions and the occurrence of dynamic regimes are described in subsection 3.2.

Steady state behavior

Stationary solutions of the model 1-9 are described in Fig.2a as the coolant temperature T_c is varied, at different effectiveness values μ_{HE} of the external heat exchanger. The reactor outlet temperature T_{out} is used as variable representative of the state of the system. At high μ_{HE}, a unique steady state solution regime is invariably observed. As μ_{HE} is decreased, two saddle node bifurcation points arise delimitating a coolant temperature range where three coexisting steady state solution regimes are detected: a low conversion, an intermediate and a high conversion solution regimes. Saddle node bifurcation points separate a range where a unique high conversion steady state solution branch is observed and a range where only low conversion steady solution regimes exist. Hence, in order to safely design and operate the monolith loop reactor, it is important to predict the evolution of these bifurcation points in the parameter space.

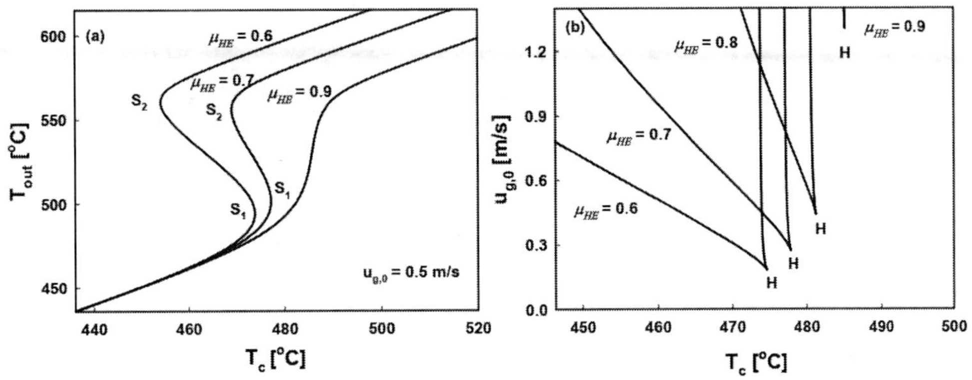

Figure 2 Steady state behavior of the monolith loop reactor; (a) steady state solution regimes as T_c varies at different μ_{HE}; (b) projections of the saddle node bifurcation points on the T_C - μ_{HE} plane at different μ_{HE} ($U_l = 1$, $Da = 520$, $B = 0.127$, $b_r = 34.91$).

Projections of the saddle node bifurcation points S_1 and S_2 on the coolant temperature T_c - reactor gas inlet velocity $u_{g,0}$ plane are displayed in Fig.2b, at different effectiveness values μ_{HE} of the external heat exchanger. It can be observed that the saddle node bifurcation points S_1 and S_2 and, hence, the region of steady state multiplicity disappear as μ_{HE} is increased. Furthermore, a limit $u_{g,0}$ value, depending on μ_{HE}, is found, under which a unique steady state solution regime is detected as T_c varies. In particular, at a given μ_{HE}, the coolant temperature range between the saddle node bifurcation points S_1 and S_2, becomes smaller and smaller as $u_{g,0}$ is decreased and eventually vanishes when the cusp point H is crossed.

Stability analysis and dynamic regimes

Results of the stability analysis of the monolith loop reactor are presented in Fig.3a. Here, steady state solution regimes and their stability characteristics are described as the coolant temperature is varied. Stationary regimes become unstable through two catastrophic Hopf bifurcation points, HB_1 and HB_2. These give rise to periodic solution branches extending over a wide coolant temperature range.

To prove it, results of reactor model simulations performed by quasi-statically varying the coolant temperature over the range delimitated by the stationary points A and B are displayed in Fig.3c and 3d. Increasing T_c starting from the stationary point A (Fig.2b), low conversion steady state solution regimes become unstable as the Hopf bifurcation point HB_1 is crossed. As result, large autonomous oscillations of the state variables arise. As the coolant temperature is further increased, stable periodic solution regimes are detected over the range 480.6 °C < T_c < 491.45 °C and suddenly extinguish at T_c = 491.45 °C. It is worth to note that autonomous oscillations are also observed when the Hopf bifurcation point HB_2 is crossed and coexist with stable stationary points in a wide coolant temperature range. A qualitatively symmetric behavior is observed as the coolant temperature is quasi-statically decreased between the stationary points B and A (Fig.3c). As the Hopf bifurcation point HB_2 is crossed, large temperature oscillations are observed. These result stable over the coolant temperature range 482.7 °C > T_c > 472.2 °C and eventually disappear at T_c = 472.2 °C.

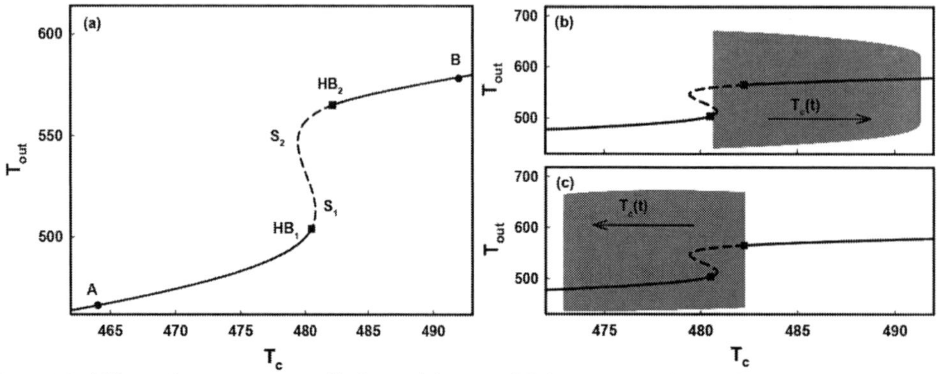

Figure 3 Stability and autonomous oscillations of the monolith loop reactor; (a) steady states and their stability (continuous and dashed lines denote stable and unstable solutions respectively); (b), (c) results of a numerical simulation performed by quasi statically increasing (b) and decreasing (c) the coolant temperature between the stationary points A and B ($U_l = 1$, $U_{g0} = 2$, $Le = 2$, $Da = 520$, $B = 0.127$, $b_r = 34.91$).

It is clear that operating conditions close to the Hopf bifurcation points HB_1 and HB_2 should be avoided in order to prevent the possibility of reactor temperature oscillations. However, because of the catastrophic nature of the Hopf bifurcation points HB_1 and HB_2, coexistence of stable periodic and stationary solution regimes is detected over wide coolant temperature ranges. When the reactor is operated in these ranges, a disturbance of the operating conditions could lead, due to a narrow basin of attraction of the desired stationary regime, to the onset of large autonomous oscillations, even when safety margins with respect to the Hopf bifurcation points HB_1 and HB_2 are kept in the parameter space. This risk is demonstrated in Fig.4, where the reactor temperature simulation response resulting from a step perturbation of the coolant temperature is displayed. The system is initially operated at the high conversion stationary solution regime corresponding to the coolant temperature value T_c = 488.376 °C. At the time t^*= 50, a step perturbation of the coolant temperature is applied. Although the final coolant temperature (T_c = 486.324 °C) is far away from the Hopf bifurcation point HB_2 (T_c = 480.22 °C), the response is oscillatory and a stable periodic solution regime is reached.

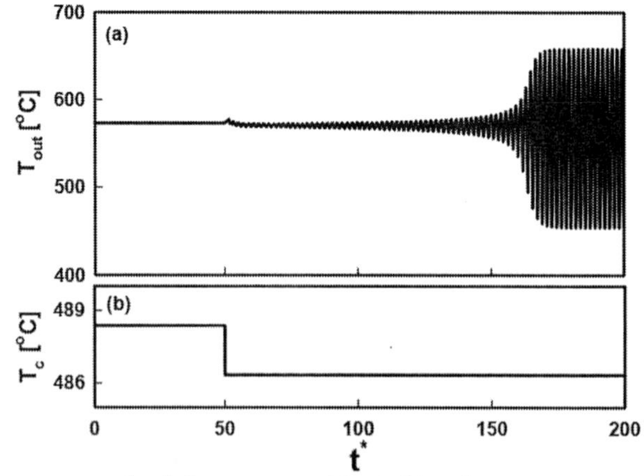

Figure 4 Reactor temperature simulation response (a) resulting from a step perturbation of the coolant temperature (b).

Conclusions

Nonlinear analysis of the monolith loop reactor for Fischer-Tropsch synthesis was performed. A mathematical model in dimensionless form was presented and bifurcation analysis was carried out examining the influence of reactor gas inlet velocity, coolant temperature and effectiveness factor of the external heat exchanger on the behavior of the system. At effectiveness factor of the external heat exchanger lower than one, two saddle bifurcation points were detected delimitating a region of the parameter space characterized by the coexistence of three stationary solution regimes. This region was found to disappear as the reactor gas inlet velocity is decreased.

Stability analysis of the stationary solution regimes of the monolith loop reactor revealed the presence of Hopf bifurcations giving rise to periodic solution branches extending over wide operating ranges. Because of the catastrophic nature of the detected Hopf bifurcations, coexistence of stable periodic solution regimes with stable stationary ones was observed. Under these conditions a slight perturbation of the operating conditions was shown to induce, due to a narrow basin of attraction of the desired stationary solution, stable autonomous oscillations with amplitude approximately equal to 200 °C ($T_{out,min}$ = 453 °C, $T_{out,max}$ = 650 °C). In view of these results, the analysis of bifurcations of periodic solution regimes of the monolith loop reactor is expected to provide significant insights to define effective strategies for safe design and control.

References

[1] R.M. De Deugd, F. Kapteijn, J.A. Moulijn, (2003), *Top. Catal.*, 26, 1-4, 29.

[2] R.M. De Deugd, R.B. Chougule, M.T. Kreutzer, F.M. Meeuse, J. Grievink, F. Kapteijn, J.A. Moulijn, (2003), *Chem.Eng. Sci.*, 58, 583.

[3] M.T. Kreutzer, P. Du, J.J. Heiszwolf, F. Kapteijn, J.A. Moulijn, (2001), *Chem.Eng. Sci.*, 56, 6015.

[4] I.C. Yates and C.N. Satterfield, (1991), *Energ.Fuel.*, 5, 168

[5] J.J. Marano, G.D. Holder, (1997), *Ind.Eng. Chem. Res.*, 36, 1895.

[6] B.B. Breman, A.A. Beenackers, E.W. Rietjens, R.J. Stege, (1994), *J. Chem. Eng.Data*, 39, 647.

33f Understanding Sulfur Poisoning in Steam Reforming of N-Hexadecane: Catalyst Characterization Studies

Satish Lakhapatri and Martin A. Abraham. Chemical and Environmental Engineering, University of Toledo, 2801 W. Bancroft, 3048 Nitschke Hall, Toledo, OH 43606

Steam reforming of logistics fuel, such as jet fuel and diesel is an effective means of hydrogen production for fuel cells. One of the main problems in reforming logistics fuels is simultaneous deactivation due to coking and presence of sulfur. Sulfur content can vary ranging from few ppm in diesel to over thousand ppm in jet fuel. In order to improve the sulfur resistance of metal supported catalysts, we must know how sulfur chemisorbs on the metal sites. Understanding the mechanism of sulfur interaction would help us get insights into deactivation and thereby help in designing a better catalyst for improved performance. In the present work, diesel steam reforming was carried out using n-hexadecane as a surrogate to study effect of sulfur and coking on deactivation.

Steam reforming of n-hexadecane was conducted at 800 deg C, 1 atm and S/C ratio of 3 with different Rh-Ni catalysts supported on γ-alumina. Thiophene was used as the source of organic sulfur at 1000 wppm. In general, the presence of sulfur increased the coking on the support but had negligible effect on the metal coke. Presence of Rh in small quantities tends to increase sulfur resistance but when present in higher amounts gave lower hydrogen yields. Lower yields, at higher Rh content, was explained based on sulfur adsorption capacities. Also, carbon morphology of spent catalysts was studied using Scanning electron microscopy. Varying degrees of both filamentous and encapsulating carbon were observed in spent catalysts. Various other factors such as nickel particle sintering, phase transformation, degree of reduction and support acidity were discussed to explain deactivation. Techniques such as X-ray diffraction, X-ray photoelectron spectroscopy, Transmission electron microscopy, Temperature programmed oxidation, reduction and ammonia desorption were employed in the present study.

172g Mathematical Modeling of Catalyst Deactivation by the Presence of Coke Formation in Catalyst Particles

M. Gaxiola-Cervantes, J. L. Rico-Cerda, and **M. Morales-Cabrera**. Chemical Engineering, Universidad Michoacana de San Nicolás de Hidalgo, Edificio M, Ciudad Universitaria, Francisco J. Mujica S/N, Col. Felicitas del Rio, Morelia, 58060, Mexico

Mathematical modeling of catalyst deactivation by the presence of coke formation

in catalyst particles

M. Gaxiola-Cervantes, J.L. Rico and M.A. Morales-Cabrera

Universidad Michoacana de San Nicolás de Hidalgo, Francisco J. Mújica S/N, Ciudad Universitaria, Col. Felicitas del Rio, México, D.F., C.P. 58060 México

Keywords: diffusion; reaction; catalyst deactivation; mathematical modelling; effectiveness factor; chemical reactors.

Abstract

The catalysts deactivation by coke deposition is a phenomenon frequently found in hydrocarbon conversion processes. Quantitative study of the activity time relation is important to determine the optimum design and operation conditions of chemical reactors [Froment, 2001].

In this work, an approximate solution for a problem of diffusion-reaction in an isothermal catalyst pellet with the presence of coke deposition is developed. The approximation is based on the linearization of the nonlinear reaction terms in the transport equations, by means a Taylor series expansion [Morales-Cabrera et al., 2005]. For the concentration profiles of all reactant species inside the catalyst particle, analytical expressions are obtained by applying the Laplace transform. As consequence, an analytical expression for the effectiveness factor, which is a function of the Thiele modulus and the time, was also derived to study the catalyst activity. And with this tool, the effect of the deactivation on the conversion in a fixed-bed reactor was studied.

The approximate analytical solution was applied to dehydrogenation of 1-buteno into butadiene and considering exponential deactivation functions to the coke deposition (Dumez y Froment, 1976), in catalyst particles with spherical geometry. The deactivation functions included dependence of the coke concentration. The predictions of the concentration profiles and the effectiveness factors were compared with results obtained by the numerical solution of the reactant transport equations and with the experimental data reported by Dumez y Froment, 1976.

AUTHOR INDEX

Abraham, Martin A. 116
Aksoylu, A.Erhan 108
Al-Dahhan, Muthanna H. 82, 107
Altimari, Pietro 109
Amlashi, Mahdi Alavi 14
Azad, Abdul-Majeed 100
Beard, Kevin D. 60
Behr, Arno 35
Bildea, Costin Sorin 109
Boghaei, Davar Mohammadi 89
Bowden, Ned B. 3
Brownscombe, Thomas 43
Brydson, Rik M. 51
Chen, Tai-Shang 80
Chern, Jia-Ming 80
Dudukovic, Milorad P. 82, 107
Espada, Loren I. 78
Ferrizz, Robert M. 78
Gaxiola-Cervantes, M. 117
Gleaves, John T. 43
Greiner, Lasse 1, 4
Grievink, Johan 109
Guo, Junxiang 84
Hackett, Simon F.J. 51
Halteman, Johnelle M. 50
Hamidzadeh, Marzieh 89
He, Lining 84
Iglesia, Enrique 99
Jiang, Xue 84, 95
Joo, Oh-Shim 76
Jung, Heon 76
Jung, Kwang-Deog 76
Kamatala, Chinna Rajanna 88
Kammler, Daniel R. 78
Khan, Jamil 26, 51
Khan, Neetha 75
Kim, Honggon 77
Kim, Hoon Sik 77
Klankermayer, Jürgen 2
Koch, Timo J. 2
Kuzeljevic, Zeljko V. 107
Lakhapatri, Satish 116
Lanterman, H. Bryan 36
Lee, Adam F. 26, 51
Lee, Hyunjoo 77
Lee, Sang Deuk 77
Lee, Sunggyu 36
Leitner, Walter 2
Li, Huaming 84, 95
Li, Hui Hui 83
Liauw, Marcel A. 4
Lindner, Falk 35
Liu, Aiguo 83
Louck, Thomas J. 78
Lu, Jidong 84, 95
Luckett, Christopher 83
Matranga, Christopher 13, 75
Monnier, J.R. 60
Monnier, John R. 81
Montoya, Joseph H. 81
Morales-Cabrera, M. 117
Mueller, Daniela H. 4
Mueller, Sean G. 82
Müller, Pia 1
Mwangi, Martin T. 3
Natesakawat, Sittichai 13
Negin, Saeedeh 89
Ojeda, Manuel 99
Picou, Jason W. 36
Pontzen, Florian 4
Putta, Satya 36
Rebelli, Jayakiran 81

AUTHOR INDEX

Ren, Junli 12
Rico-Cerda, J. L. 117
Ritchey, M. Barry 78
Rowshanzamir, Soosan 14
Runge, M. Brett 3
Sadhukhan, Jhuma 67
Sanders, Tom 13
Sapkota, Prabal 77
Schaal, Melanie T. 81
Selen-Çaglayan, Burcu 108
Snow, Clark S. 78
Song, Kwang Ho 77
Styring, Peter 26, 51
Sun, Runcang 12
Sundararajan, Desikan 100
Turbeville, Wayne 83
Utkoor, Umesh Kumar 88
Valipour, Parisa 14
Van Ommen, J. Ruud 109
Van Zee, J.W. 60
Veser, Goetz 13
Wenzel, Jonathan E. 36
Williams, Christopher T. 81
Wilson, Karen 51, 67
Xia, Jiexiang 95
Xu, Dan 95
Xu, Feng 12
Xu, Mian 67
Yablonsky, Gregory S. 43
Yan, Yongsheng 84, 95
Yang, Dae Ryook 76
Zhang, Yi 13
Zheng, Xiaolin 43
Zhu, Wenshuai 84, 95

American Institute of Chemical Engineers
3 Park Avenue
New York, New York 10016-5991

ISBN: 978-1-60560-223-3